The
OLMSTED PARKS
of Louisville
of Louisville

Happy Botanizing!
Patricia Dalton Haragan
March 26, 2014

The OLMSTED PARKS of Louisville

A BOTANICAL FIELD GUIDE

Patricia Dalton Haragan

Photographs by
Susan Wilson and Chris Bidwell

UNIVERSITY PRESS OF KENTUCKY

The financial support of the Kentucky Society of Natural History is gratefully acknowledged.

Scholarly publisher for the Commonwealth,
serving Bellarmine University, Berea College, Centre
College of Kentucky, Eastern Kentucky University,
The Filson Historical Society, Georgetown College,
Kentucky Historical Society, Kentucky State University,
Morehead State University, Murray State University,
Northern Kentucky University, Transylvania University,
University of Kentucky, University of Louisville,
and Western Kentucky University.
All rights reserved.

Editorial and Sales Offices: The University Press of Kentucky
663 South Limestone Street, Lexington, Kentucky 40508-4008
www.kentuckypress.com

Library of Congress Cataloging-in-Publication Data
Haragan, Patricia Dalton, 1953–
    The Olmsted parks of Louisville : a botanical field guide / Patricia Dalton Haragan ;
photographs by Susan Wilson and Chris Bidwell.
        pages cm
    Includes bibliographical references and index.
    ISBN 978-0-8131-4454-2 (pbk. : alk. paper) — ISBN 978-0-8131-4456-6 (pdf) —
ISBN 978-0-8131-4455-9 (epub)   1. Plants—Kentucky—Louisville—Identification.
2. Plants—Kentucky—Louisville—Pictorial works.   3. Parks—Kentucky—Louisville.
I. Title.
    QK162.H37 2014
    581.09769'44—dc23
                                                                2013039723

Manufactured in Canada

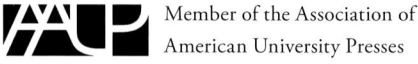
Member of the Association of
American University Presses

To my daughters, Molly and Anna,
a privilege and joy.
With love.

In loving memory of our dear
friend Vincent Nold, our
"brother" and sunshine.

Contents

Foreword: Reading a Place *Daniel H. Jones* ix

Introduction *Susan M. Rademacher* 1

About This Book 7

Illustrated Plant Structures 21

Plant Descriptions of 384 Species 31

Ferns and Fern Allies *33*

Herbaceous Plants, Woody Vines, and Shrubs *42*
Arranged by Flower Color, Season, and Alphabetically
by Family within Season
Spring: February, March, April
Summer: May, June, July
Fall: August, September, October

White Flowers *42*

Yellow/Orange Flowers *140*

Red/Pink Flowers *202*

Blue/Purple Flowers *232*

Green Flowers *287*

Brown Flowers *328*

Trees *335*

> Arranged by
>> Evergreen: Needlelike or Scalelike
>> Deciduous: Leaves Alternate or Opposite, Simple or Compound,
>> and Alphabetically by Family within Group

Sedges, Rushes, and Grasses *388*

Glossary   409

References   419

Acknowledgments   423

Olmsted Parks Conservancy   425

Index   427

Foreword

Reading a Place

Stories fill our landscapes—from the densest urban core to the wilds of Montana—but finding and reading those stories requires time and some acquaintance with the special places of a site and its inhabitants. In this regard, a park is no different from Paris: to know it, you must spend time there. And the magic of Louisville's Olmsted Parks is that they are gateways to some remarkable places, located just a few minutes down the road from every neighborhood in the city. A glance at a Louisville map shows immediately the genius of the design: from Shawnee and Chickasaw Parks in the floodplain of the Ohio River, to Iroquois in the knobs south of town, to Seneca and Cherokee in the eastern limestone uplands, the system reflects the geographic and natural diversity of Louisville.

To unravel the stories hidden in these landscapes requires knowledge, and this is where Pat Haragan's *The Olmsted Parks of Louisville: A Botanical Field Guide* becomes your essential companion. Reading a place, as reading a book, demands a vocabulary, and a good field guide is like a good dictionary: learn the plants and you will learn to read the place, as each species occupies a niche shaped by geology, soils, past disturbance, and interrelations with other species. If you carry this guide with you and slowly master the language, new perspectives open. The uniqueness of the temperate deciduous forest of eastern North America with its oaks, migrating songbirds, and spring wildflowers contrasts sharply with the prairies to our west, the boreal forests to our north, and the subtropics to our south. As you learn the lexicon of plants, dialects emerge: the chestnut oaks of the Iroquois hills are distinct from the chinkapin oaks of Cherokee Park or the cottonwoods of the Shawnee floodplain, and there are reasons for this rooted in the soils, rocks, hydrology, and human history. New conversations become possible: the past uses of plants by Native Americans, a plant's edible characteristics, their relations with distant cousins in the tropics or the arctic.

This is how we come to know place, and how we give identity to that distinctive Kentucky landscape that is essential to our city, and to ourselves. Our rolling limestone hills, their amazing fossils, spring colors, and the diverse wildlife that in turn feeds on those plants, encompass our natural history. While a field guide begins as a dictionary, it becomes a tour book, introducing a world of stories that is limited only by your willingness to spend time unraveling them. Somewhere in that journey, the deeper magic of our parks will reveal itself. Without such a guide, city dwellers will never come to know nature or develop a deeper understanding of place, and that is a loss. Build them and love them, and suddenly you connect to the ancient world and rhythms that linger behind our buildings and beneath our pavements. The twenty-first century will be an urban century, and the ability to make those connections is no small thing, and books like this will guide those urban billions back to nature.

DANIEL H. JONES
Chairman and CEO, 21st Century Parks, Inc.

The
OLMSTED PARKS
of Louisville

Ohio River

Shawnee Park*

Northwestern Parkway

Chickasaw Park

Southwestern Parkway

Market Street

Broadway

Greenwood Ave

Magazine Street

Cypress St

264

Boone Square*

Elliott Square

Victory Park

20th St

28th St

22nd St

Bank Street

Main St
Market St

Baxter Square

Third Street

65

Broadway

Hancock St

Shelby Street

Third Street

Fourth Street

Kennedy Street

Park Ave

Central Park

Algonquin Park

Algonquin Parkway

University of Louisville

Stansbury Park

Byrne Ave

Floyd Street

Third Street

Oakdale Ave

Wayside Park

Churchill Downs

Southern Parkway

South Third Street

New Cut Road

Taylor Boulevard

Dixie Highway

264

Iroquois Park*

Marshall Road

River Road

71

Canal Ave

Bingham Park

Brownsboro Road

Frankfort Avenue

Cannons Lane

Cherokee Park*

Cherokee Parkway

64

Pee Wee Reese Rd

Lexington Road

Seneca Park

Bowman Field

Taylorsville Road

64

Baxter Avenue

Grinstead Drive

Willow Park

Tyler Park Dr

Tyler Park

Barret Avenue

Shelby Park

Eastern Parkway

Bardstown Road

Churchill Park

Preston Highway

Crittenden Drive

Kentucky Fair and Exposition Center

Louisville Zoo

Henry Watterson Expressway

264

65

Louisville International Airport

Technology Park

N
W E
S

* Park designed by Frederick Law Olmsted, Sr.
in collaboration with his partners before his retirement in 1895.

Copyrighted mapping data
provided by LOJIC

LOJIC

Miles

1   2   3   4   5

Introduction

SUSAN M. RADEMACHER

Parks Curator, Pittsburgh Parks Conservancy
Founding Executive Director, Louisville Olmsted Parks Conservancy

Who was Frederick Law Olmsted? And what is special about his Louisville parks as containers for such a stunning array of native and naturalized plants?

This towering shaper of the American landscape explored many roles— seaman, farmer, mine manager, abolitionist, Federal Department of Sanitation Commissioner, writer, and publisher—on his journey to becoming the nation's first landscape architect. As befits a pioneer, Olmsted was obstinate and strong willed, true to his convictions. He was labeled "long-headed" for his obsession with planning and his ability to project his ideas for generations into the future.

At age twenty-eight, on a life-changing tour of England, Olmsted was thrilled to experience a landscape designed as a work of art. His response upon visiting Capability Brown's Eaton Hall was to write: "What artist, so noble, has often been my thought, as he, who with far-reaching conception of beauty and designing power, sketches the outline, writes the colours, and directs the shadows of a picture so great that Nature shall be employed upon it for generations, before the work he has arranged for her shall realize his intentions?"

Olmsted well understood that his work was a pact with nature. The core tenets of his philosophy were these:

- The experience of nature (scenery) should be available to all people.
- Contact with nature has the power to soothe human souls and promote health.
- Experiencing nature's beauty promotes moral perception and intuition.

Olmsted was born in 1822 in Connecticut, and his love of scenic vistas and outdoor skills were cultivated on frequent rural jaunts with his father. John

Olmsted was an avid reader of Humphrey Repton and J. C. Loudon, the connoisseurs of the picturesque, the beautiful, and the sublime in the landscape. As a young man, his eye was trained on those qualities, whether he was about the business of farming, mining, or journalism. In England, he realized, "Beauty, grandeur, impressiveness in any way, from scenery, is not often to be found in a few prominent, distinguishable features, but in the manner and the unobserved materials with which these are connected and combined. Clouds, lights, states of the atmosphere and circumstances that we can not always detect, affect all landscapes. . . . Gradually and silently the charm comes over us; the beauty has entered our souls; we know not exactly when or how." When he finally came into his own as a park designer, Olmsted worked hard to provide opportunities for city people to experience such feelings and benefit from their restorative power. In Louisville, near the end of his career and perhaps more than anywhere else, Olmsted was given the perfect conditions to do just this.

Louisville's parks system is a homegrown idea. It was born of civic leadership among members of the Salmagundi, an all-male social and literary club founded in 1879 (and still active today). In 1887, the Salmagundi joined with the Commercial Club of Louisville to sponsor studies and create legislation to establish the Louisville parks system. Over 1,000 acres were immediately purchased by the new Park Board, who in 1891 invited Olmsted to review their ideas for a park system. Olmsted was promptly commissioned to produce plans for the new "necklace of parks" and parkways. The result was one of only five such systems in the country, representing the last commission of this type in the elder Olmsted's career.

By 1891, Olmsted had been practicing as a landscape architect for thirty-four years and had defined the profession in the course of making his first great work, with the architect Calvert Vaux: Manhattan's Central Park. With a love of nature instilled from childhood, a background in farming and journalism, an ardent social conscience, and the ability to take the long view, Olmsted was superbly equipped to shape the American landscape, from the design of communities, campuses, and estates, to the creation of parks.

But Kentucky already had a special appeal to Olmsted. In 1853, during his third trip through the South to study slave-based agriculture, Olmsted journeyed by stagecoach from Cincinnati to Lexington, and then by train from Lexington to Louisville. So struck by the landscape, he declared Kentucky to be the most beautiful natural parkland he had ever seen.

As the ultimate park system of Olmsted's career, the Louisville parks reflect his mature vision. He wished each park to fulfill a specific function for the whole city—and Louisville's topographical variety made this task easy. With Shawnee's flat Ohio river terraces, Cherokee's rolling Beargrass Creek valley, and Iroquois' rugged "knob" of old-growth forest, the trio of landscapes complement each other as components of one great urban park. Totaling 1,200 acres, the parks were designed to offer a complete range of park experiences as defined by Olmsted, from civic gatherings and social interactions to organized athletics and personal recreation. The parks' west, south, and east locations anchor each distinctive region of the city, and the parkways connect people throughout the city to the Kentucky landscape.

Olmsted's 1891 report to the Park Commission analyzed the unique natural character of each park and recommended that each should be designed to reflect the unique character of its site, with none requiring much manipulation of the land. Shawnee Park's beautiful riverside expanse was well suited to accommodate crowds of people brought together for a great variety of activities and entertainments on broad swaths of lawn, in picnic groves, and around flower gardens. The terraced landscape reaching down to the river's edge was the only area to appear naturalistic, the effect of which was largely created through extensive plantings. Wildness came into Shawnee in bits and pieces over the ensuing decades, through a gradual lessening of intensive maintenance controls and the ripe environment for naturalizing plant species that river corridors present.

Cherokee Park, considered Olmsted's most scenic work, focuses on the inner landscape experience of traversing from ridgeline to stream valley. As he said, "If you want the refreshment that is to be had in the contemplation of superb umbrageous trees, standing singly and in open groups distributed naturally upon a gracefully undulating greensward, to procure such scenery in higher perfection that, with large outlays to obtain it, is yet to be found in any public park in America, all that is needed is the removal of fences and a little judicious use of the ax on your Cherokee Park site." This was to be the "pastoral haven" of greensward, groves, and water that was Olmsted's ideal.

About Iroquois, he said, "If you want as a treasure of sylvan scenery . . . the grandeur of forest depths in the dim seclusion of which you may wander musingly for hours, this you may find ready to your hand on the Iroquois Hill." This was to be the scenic reservation—a wilderness experience of wild slopes blanketed with deep forest and breathtaking savannah and summit views to the surrounding countryside.

Over time, of course, nature has worked it out with Olmsted. Their collaboration has been challenged and compromised by development, fire, overuse, exotic horticulture, and critters of all kinds. Recognizing that the parks' value had eroded, a renaissance of reeducation and reinvestment began in the 1980s and has continued apace. The central question is this: How to introduce new and evolving uses without destroying the Olmsted legacy? Officially adopted by the City of Louisville in 1994, the *Master Plan for Renewing Louisville's Olmsted Parks and Parkways* is notable for its interdisciplinary approach to restoring both the cultural and the ecological landscape, on a foundation of historic research, public participation, and skilled maintenance/management.

The Olmsted Parks must be restored, preserved, and enhanced to continue their enormous contribution to the quality of life in Louisville. They are an incomparable gift from a remarkable civic partnership that, a century ago, championed planning, raised substantial money, and summoned the goodwill and resources of the community at large. As in most cities, the Olmsted parks and parkways were just the beginning. As the parks continue to grow, new plans must seek the optimum balance of natural systems, historic values, and use and management. Designed to bring people into transformative contact with nature, the Olmsted parks can be viewed as the source of inspiration for parks of the future, as well as a source of inspiration in everyday life.

The beauties of these Olmsted parks have inspired generations of painters, poets, and just plain folks with an ever-changing panorama of ephemeral effects, as is so well documented in this significant volume by the superb botanist Pat Haragan. For Louisvillians, the parks are touchstones and points of reference, inexorably tied to who we are. It may be hard for us in our world today—so full of images and patterns and visual excitement that call attention to themselves—to understand and appreciate a design ethic that is soft-spoken and subtly affecting. That was Olmsted's ethic, and it is the ethic we must invoke in the continued effort to preserve the native character of these wondrous landscapes while respecting them as works of art in nature.

Trout Lily

You see me as tedious,
the small print easy to overlook,
your attention always elsewhere
on the ash and sycamore,
or that great rolling belly of meadow,
sliced through by the creek.

The world's peppered with grand passions:
Big cars. Houses with wide mouths.
Everything glitter and volume.

But don't take my delicacy
for lack of passion. It's just my desire
for drama on a small scale:

pulse of light and shadow,
the patience of waiting out the seasons
for my moment of truth
so I can put on this gold face.

All this push and shove
to surge toward the light.
And in the end, the welcome turning—
tug of earth, the quiet exit.

Paula Keppie
Louisville Review, *Spring 1996*

About This Book

PATRICIA DALTON HARAGAN

The purpose of this local field guide is to aid in the identification of and spark an interest in the 384 plants highlighted herein, which grow within the five most popular Olmsted Parks: Cherokee, Seneca, Iroquois, Shawnee, and Chickasaw. Located within the city limits, these signature parks together total roughly 1,990 acres. Although this book in no way contains every plant found in these parks, it does consist of a broad sampling of species that may be encountered while hiking the many trails and walking paths. Trees, shrubs, vines, wildflowers, ferns, invasives, wetland plants, weeds, and grasses are included. Learning how to identify and name the plants growing in one's local surroundings forms the basis for our "roots," our sense of place and what we learn to value. This, in turn, leads to a greater awareness of and appreciation for the plants in each of these unique urban green spaces as well as provides hours of outdoor enjoyment. The rich history of these beloved Olmsted Parks defines us as Louisvillians: it is part of our heritage, our past. My goal is to reward the reader with the thrill of discovering the fascinating world of plants, and I hope that this in turn will stimulate interest in future studies to document the flora in conjunction with preserving the artistic legacy of this great landscape for generations to come.

How to Use This Guide

This book is arranged in four parts. (1) Fern and fern allies (seedless vascular plants) are presented in the first small section of the book. (2) The second section contains the bulk of the species, or those with recognizable flowers. Here are found the herbaceous plants, woody vines, and shrubs. These are arranged by flower color, then flowering season, and within season, alphabetically by family.

It must be noted that color and season often are not clear-cut categories. For example, many flowers given in a color section have strong tinges of another color. If a plant is not found in one section, the reader is urged to look for it in a closely related color category, and likewise with seasons. (3) The third section consists of trees. Because their flowers are often inconspicuous, they are organized by leaf type, first separating the evergreen conifers from the deciduous broadleaf trees. The latter much larger group is further divided first by leaf arrangement and then by leaves—simple or compound—and alphabetically by family within the groups. (4) The last section contains the grasses, sedges, and rushes and is arranged alphabetically by scientific name. Many of the entries in this book appear on a single page and consist of a detailed description written with the nonprofessional botanist in mind. For an easier approach, go directly to the "Key Features" heading. It contains a few characters that highlight the species for quick identification. However, if more than one species is included on the page, then only a few distinguishing characteristics will be given for the second entry, with contrasting characters in bold print. Most of the species are coupled with color photographs that enhance and magnify the detailed descriptions. One goal of this book is that the user will "grow" into this field guide. It is in no way a technical manual, although some technical terms exist in the descriptions. Over time, and with use, the reader will eventually learn the meaning of a "disk floret" or "obovate" leaf blade. This way the reader can keep adding to his or her botanical knowledge by simple observation. Using and understanding the technical language also adds consistency and precision to the task of describing species and learning to separate one from another.

The entries in each of the four sections all follow the same format. At the very top of the page are abbreviations for the parks:

Cher = Cherokee Park
Sen = Seneca Park
Iroq = Iroquois Park
Shaw = Shawnee Park
Chick = Chickasaw Park

This information will help the reader when exploring the different parks. Many species are found throughout all the parks, and some only in one or two. The common name of each plant and its family are followed by its scientific name. I have adhered to scientific names used in the state's only comprehensive botanical manual, *Plant Life of Kentucky: An Illustrated Guide to the Vascular Flora* (Jones 2005). This work has become the standard botanical text for the state and provides a guide for consistency in this current work. The symbol *"syn,"* for synonym, refers to an older name that is no longer recognized but at one time was popular. Older botanical guides may have used the synonym listed.

In some entries, the term **Invasive plant** is used. This means that the plant has been introduced into an environment in which it did not evolve and has very few natural enemies, if any, to keep it in check. It is highly aggressive and adaptable, and it has the ability to reproduce quickly and expand its numbers and coverage rapidly, thus invading new habitats. I have also included the status of each given by the Kentucky Exotic Pest Plant Council (www.se-eppc.org/Ky/), which is a reference that provides the latest information, with maps, on the spread of these species in the state and beyond (KEPPC 2012).

In bold print are the headings used in each description. All are self-explanatory and neatly separate one character from another. The distribution contains habitat information as well as frequency of the species. Using **abundant, common, uncommon,** and **rare** is subjective, but I use the following definitions in this book. **Abundant:** meaning that you will find this plant in large quantities throughout a given area. **Common:** that you will probably find the plant in a particular area in great numbers. **Uncommon:** that a species may not be found in a particular habitat or may occur in small numbers. **Rare:** that a plant has a very limited number in the habitat, ranging from one to few. I have tried my best to decipher the status of each species by relying on my own field observations as well as those of others who know the park flora.

The last heading, "In Kentucky," refers to the three major physiographic provinces found in the state: Mississippi Embayment (ME), Interior Low Plateaus (IP), and Appalachian Plateaus (AP) (Figure 1, modified from Jones 2005). Each region is unique, with the underlying geologic parent material influencing plant diversity within the area. I have included this additional information for each species to give the reader an idea of the general geographical range of each species. If a species is widespread throughout the state then AP, IP, ME are listed. Others might have a more specialized distribution, such as IP or ME. This information follows the online *Atlas of Vascular Plants in Kentucky: a First Approximation* (J.J.N. Campbell and M.E. Medley 2012) at http://bluegrasswoodland.com, with current data on plant species taken from herbaria throughout Kentucky and the United States.

The ending paragraphs include folklore, origin of common and scientific names, medicinal uses, life histories, ecological or ethnobotanical details, and other information that might be of interest to the reader.

Most of the ethnobotany highlighted in this book is taken from Moerman (1998) and is included only for historical background and for general interest.

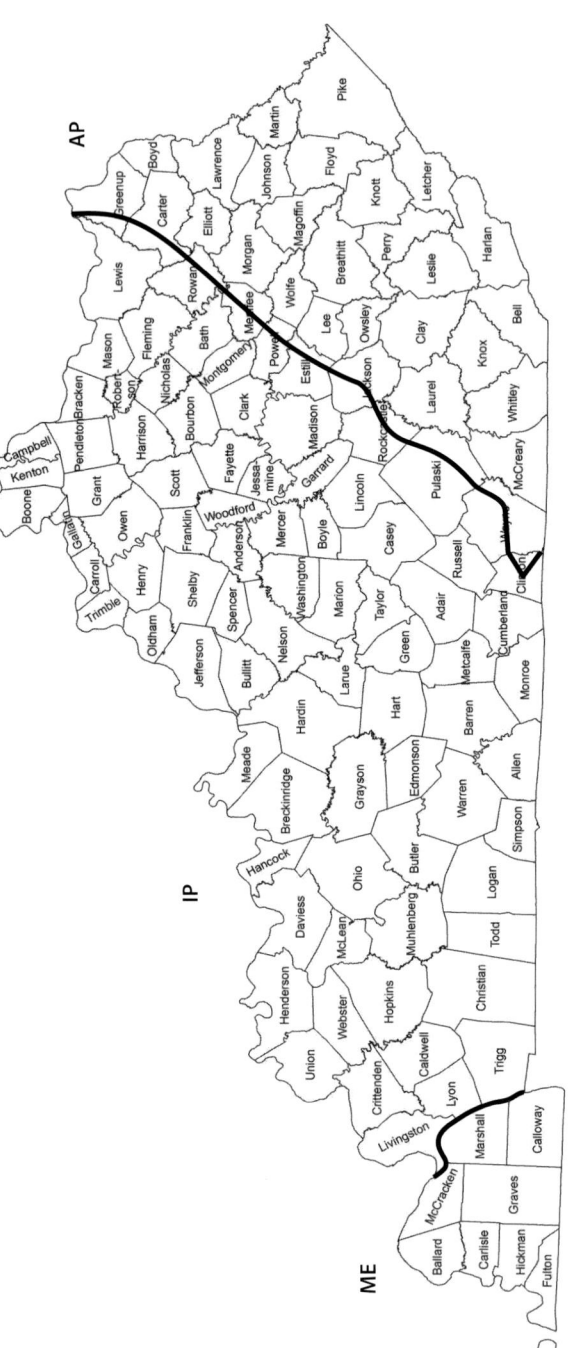

Physiographic provinces of Kentucky: Mississippi Embayment (ME), Interior Low Plateaus (IP), and Appalachian Plateaus (AP).

It must be emphasized that the information related to the use of wild plants for foods and medicines in this book is provided only as a general guide. Some people may experience harmful reactions to wild plant foods and medicines that are generally regarded as safe for most people. In particular, the information in this book should not be used for self-medication without consulting a physician—this can be very dangerous. Attempts to utilize wild plants in any way depend on many factors controllable only by the reader, and the author and publisher assume no responsibility in the case of adverse effects in individual cases (modified from Jones 2005).

Brief Overview of the Five Parks

CHEROKEE PARK

Cherokee Park, originally named Beargrass Park, opened in 1892. Today, it is considered the most popular urban park in the state and is also ranked in the top 100 parks in the country. Encompassing 389 acres, the land is made up of a mix of gently rolling to steep hills, limestone outcrops, open vistas, and woodlands that are bisected by the scenic Middle Fork of Beargrass Creek, which meanders on its way to the Ohio River. In addition, the park also contains a golf course, playgrounds, picnic areas, sports fields, hiking, and biking trails. The Scenic Loop is a 2.4-mile recreational path and park drive that follows the creek, crisscrossing back and forth over historic stone bridges.

This parcel of land was once made up of a towering canopy of oaks, hickories, chestnuts, beech, tuliptrees, basswoods, Ohio buckeyes, and other species that define the region known as the western mixed mesophytic forests. The understory contained a few shrub species, dense patches of cane, and a rich and diverse herbaceous layer. Underlain by limestone, the land was fertile and abundant with natural springs, creeks, caves, and sinkholes. By the mid-1800s, as settlers moved into this hilly terrain, the trees were cut down and the forests were cleared, mostly for grazing livestock. This destruction of the forests transformed the landscape, and the deterioration of the forest continues today with the onslaught of extreme hot summer temperatures, drought, wind and ice storms, and invasion by exotic plant pests and disease.

In 1773, Cherokee was originally part of a 4,000-acre military land grant that was available to help military spouses and families settle in the area, but it wasn't until 1893 that the park started to take its present-day shape. Six major estates with land holdings belonging to the families of Cochran, Bonnycastle,

Barrett, Alexander, and Morton and Griswold were purchased by the Board of Park Commissioners. Later, smaller lots were donated, bought, and pieced together to make up the acreage as we know it today.

Frederick Law Olmsted Sr. designed Cherokee with the goal of working with the natural landscape. The open, gently rolling hills and the pockets of dense forests on the steep slopes and along the creek running through the valley bottom were the perfect slate for his 1891, 1894, and 1897 plans using both native and nonnative species. Today, it is almost impossible to know if native plant material was part of the original Olmsted plantings or growing naturally.

In 1974, a major catastrophe struck when a severe tornado touched down along Eastern Parkway and raged through Cherokee Park. In a matter of 20 minutes, over 2,000 trees in the park had been leveled. The devastation to the health of the park, the park budget, and the hearts of the people was unimaginable. The closed canopy of trees now became open disturbed ground, prime areas for invasive plant species and diseases to take hold and quickly become established. This natural disturbance, coupled with human disturbance and lack of adequate funding for management practices over the years, has taken a major toll on the park's integrity. The alarming deterioration of the woodlands was addressed with detailed recommendations for ecological restoration by the 1994 *Master Plan for Renewing Louisville's Olmsted Parks and Parkways* produced by Andropogon Associates, and others that made up the Master Planning team. This was followed by the Olmsted Parks Conservancy's launch of a much-needed volunteer-based restoration program. In 2005, the Conservancy undertook the Woodlands Restoration Campaign, whose mission is to obtain funding to help "restore, enhance and preserve" Louisville's historic Olmsted Parks. One major goal is to try to regain ecosystem stability with an emphasis on invasive plant removal using sustainable management practices.

SENECA PARK

Named after the Seneca Indians, Seneca Park adjoins Cherokee via Park Boundary Road and was the last park designed in Louisville by the Olmsted Brothers Firm in 1928. Consisting of 526 acres (including land used for Bowman Air Field), this manicured park is popular because of the 18-hole golf course and various sports facilities, including tennis and basketball courts, a soccer field, and a 1.2-mile walking track. Its formalistic style complements the naturalistic design of Cherokee. Like Cherokee, Seneca was once extensively covered by a forest of similar composition. Today, only a few small woodland fragments are left by the golf course and near scenic Beargrass Creek. They have been heavily disturbed by foot traffic, cyclists, and natural disturbance, so native shrubs and wildflowers are scarce.

This land, made up of gently rolling hills and mossy limestone ledges bordering Beargrass Creek, was once the property of James D. Breckinridge (1781–1849), a prominent lawyer and U.S. Representative. Eventually, the land was passed on to the Baron Waldemar Konrad von Zedtwitz, a German national, through his American mother, Mary Elizabeth Breckinridge Caldwell. During World War I, it was seized by the United States government as enemy property, but the baron was able to reclaim his mother's estate, and in 1928 the land was sold to the Board of Park Commissioners. Soon after, the design plan for Seneca Park began (Louisville Friends of Olmsted Parks, 1988). Today, a natural spring, called Breckenridge Spring or Bowman Spring, is located on the southeast section of the park close to Bowman Air Field. A beautiful, architectural stone wall surrounds the springhead, where cool waters flow westward throughout the year. In late summer and fall, colorful wildflowers adorn the creek's edge. There is an ongoing restoration project to clear the invasive plants from the spring and surrounding woods. When completed, this unique natural feature will be a peaceful asset for the public to enjoy.

On an expansive hill hemmed in by Interstate 64 and Seneca Park Road is an area known as "Cedar Hill," which contains a special collection of Eastern red-cedars started by Michael Hayman, the arborist for the City of Seneca Gardens. These locally collected, grafted trees are being grown to see which have potentially useful ornamental value (such as variation in shape and color). Most come from the knobs of Indiana; others were collected south of Louisville and around Frankfort, Kentucky. This hill overlooks a gently winding stretch of Beargrass Creek as it flows into Cherokee Park and offers a perfect spot for repose.

IROQUOIS PARK

This magnificent park, once regarded as "Louisville's Yellowstone," was designed in 1897 by Olmsted and consists of 725 acres located at the foot of Southern Parkway, south of downtown Louisville. Built on a large forested cone-shaped "knob" with rugged terrain of unparalleled beauty, Olmsted's plan for this park was "to design a scenic reservation with very little forest alteration." He intended to provide the visitor with a sense of contact with a native forest common to the region and complete separation from the expanding urban surroundings. Even to this day, when walking the hiking or bridle trails that traverse the park, one has a sense of being immersed in total wilderness.

Olmsted envisioned Iroquois as "a treasure of sylvan scenery," and there are so many attractions to this park that the visitor will find plenty to do. Besides reveling in the natural beauty, there is an open-air amphitheatre for local performances, playgrounds, an 18-hole golf course, and picnic areas. Uppill Road winds through the park and ends at the top of the steep landform formerly called "Burnt Knob." The flat 45-acre summit, called "Summit Field," is the park's highest point (761 feet). Its oak savanna, an open landscape dotted by a few trees, was originally created and maintained by lightning strikes and grazing of wild animals. Later, the Native Americans kept the oak savanna open by repeatedly setting fires; this management practice was reinstated after restoration of the savanna in 1996 with periodic controlled burns. Today, this restored open oak savanna with prairie grasses and wildflowers is once again a haven for wildlife. Mown paths lead through this spectacular feature where several ponds, benches, and a picnic pavilion reside. In addition, four vistas—South Lookout or Panther Point, West Lookout or Observation Point, Krupps Point, and North Lookout—provide stunning panoramic views of the surrounding area, including downtown Louisville.

Although more fragmented today, the forest cover as described in the 1994 Master Plan by Eco-Tech, Inc., still consists of uneven-aged, second-growth forests interspersed with some remarkable stands of magnificent large oak, beech, and maple. The presence of fallen logs and dead snags suggest that natural processes of tree gap regeneration have occurred over time (Andropogon Associates 1994). The bedrock is made up of New Albany shale of Devonian age covered by a mantel of windblown loess soil that originated from soils far to the north exposed by receding glaciers thousands of years ago. This silty loess deposit that covers the knobs and slopes is highly erodible. Steep hillsides through the park have deep gullies where the soil has washed away, especially along the hiking and rogue trails. Nonnative plants have made inroads into the woodlands through these disturbed hillsides.

Iroquois is remarkable because of its diverse plant communities, starting with the stunted, gnarly blackjack oak-post oak forests just below Summit Field on the southern and southwestern steep slopes with poor dry soils and transitioning through the chestnut oak woodlands on many of the upper and middle slopes. Many mature stately chestnut oaks with deeply furrowed bark stand out in these woodlands, especially on the south-facing slopes. The sparse understory here includes beautiful spring-flowering downy serviceberry trees, low-bush blueberries, and a few herbaceous species. On the middle and lower slopes, the oak-hickory woodlands prevail and are made up of such dominant species as northern red oak, chestnut oak, white oak, shagbark hickory, and pignut hickory.

In the low, moist woods on the southeast side bordering Palatka Road is the towering beech-tuliptree-sweetgum woodlands. This latter community has the richest herbaceous flora in the park, with many common and rare species growing in this area. Beginning in March, wildflowers and ferns such as white fawn-lily, wild comfrey, colonial dwarf dandelion, dwarf crested iris, five-parted toothwort, Virginia spring-beauty, star chickweed, southern adder's-tongue, southern beech fern, and many more are spotlighted by the dappled spring light while one hikes the horse trail throughout this grand section of Iroquois.

SHAWNEE PARK

Consisting of 284 acres, Shawnee Park is bordered by the Ohio River to the north and Southwestern Parkway to the south. A paved, multiuse River Walk path parallels the river and extends eastward. This seven-mile path is popular with hikers and bikers as they travel through the upper floodplain woodlands. This area was heavily disturbed by previous grazing of livestock and by human activity, and floodwaters regularly ravage these woodlands, where now mostly nonnative species prevail. A few remnants of these floodplains woodlands can be seen in the spectacular mature cottonwood trees growing to 100 feet or more. These mighty specimens tower above the other accompanying trees—such as silver maples, boxelders, white and blue ashes, hackberrys, wild cherries, and sycamores—and their presence enhances the beauty of this powerful river as it flows downstream into the Mississippi River. Spring wildflowers are rare, but a few, such as Virginia spring-beauty, white fawn-lily, and five-parted toothwort, can be found along the north-facing disturbed open woods above the River Walk, where invasive ground covers and shrubs dominate.

Shawnee was once homesteaded by three families, with orchards of fruit and nut trees planted in the broad open spaces with a few scattered groves of trees. This was the perfect canvas for Olmsted's design in 1893. He envisioned "broad and tranquil meadowy spaces" with a central great lawn to be the core of this park, which is dotted with picnic areas and playing fields and offers views of the Ohio River from the terraced landscape.

Located off Southwestern Parkway in Louisville's west end, this 61-acre park just west of Shawnee was created in 1921 by the Frederick Law Olmsted's successor firm, the Olmsted Brothers. This pastoral land, which overlooks the Ohio River, was previously the country estate of politician John Henry Whallen. The design for this park was formal with an emphasis on space for "active recreation." There is a one-mile-long walking path that winds through the expansive mown areas and small woodlands, a scenic fishing lake (catch and release), picnic shelters and tables, grills, horseshoe pits, and six clay tennis courts.

In the second decade of the twentieth century, racial tension and the segregation movement against African Americans were peaking. Bowing to pressure, the Park Commissioners passed a resolution segregating all the parks and designating Chickasaw and four other city parks for use by African Americans only. They did hire the Olmsted Brothers to design two of them—Chickasaw and Algonquin Parks. Olmsted Brothers could not change this new policy, which the senior Olmsted would have despised, but they made plans for well-designed landscapes with all the amenities. Frederick Law Olmsted Sr. was a Social Democrat and fervently believed that parks should expand personal freedom and be made available to people of all walks of life. It wasn't until 1954 that Mayor Andrew Broaddus banned segregation, saying that the parks are for all people, in agreement with the socially progressive views of Frederick Law Olmsted Sr. (Louisville Friends of Olmsted Parks 1988).

TRANSPORTED LANDSCAPES

For centuries, people have been rearranging the world's flora through accidental or intentional introductions. Floras published in the nineteenth century that offer a glimpse of the native and transported landscape in the Louisville area are rare. However, one very informative book written by H. McMurtrie and published in 1819, *Sketches of Louisville and its Environs; Florula Louisvillensis*, included 400 genera and 600 species of plants. One of Kentucky's most brilliant and eccentric botanists, Constantine Samuel Rafinesque, traveled through Louisville in 1818, combing the area for plants that were new to science and amassing a large collection, before he secured a teaching position in Lexington at Transylvania College. Since he was a zealous collector of natural history, his early botanical specimens from this time are presumed to be important contributions to the plant data in McMurtrie's book, although no credit was given to Rafinesque.

McMurtrie states that at the time of settlement, from 1773 to 1784, this "Western Territory" now known as the Ohio Valley was rich in wildlife—elk, deer, otter, bear, wolves, and buffalo were plentiful—and that the land was covered in extensive wetlands and vast, dense forests. Magnificent native trees such as walnut, red mulberry, locust, beech, sugar maple, cherry, pawpaw, buckeye, elm, tulip-poplar, dogwood, oak, and hickory thrived. Many trees measured 6 to 10 feet in diameter, and the tulip-poplars were said to be gigantic in comparison, often 22 feet in circumference. In the fertile and wealthy Bear-grass settlement, massive native grape vines with a circumference of 36 inches were not uncommon. However, in addition to this rich, diverse native local flora, the early settlers brought with them plants from their homeland, which included belladonna, onion, field garlic, teasel, chamomile, oats, celandine, mugwort, black mustard, poison hemlock, ox-eye daisy, European privet, common millefoil, ground-ivy, English plantain, white and red clover, sow-thistle, Timothy grass, wild carrot, and catnip, to name a few. Many of these transported plants, mostly of European origin, were purposely introduced and are still with us today, growing in our yards, gardens, fields, and woodlands.

Frederick Law Olmsted used both native and nonnative plants, mostly of European and Asian origin, in his design plans to "evoke a range of landscape characters." His designs included a rich herbaceous layer, an understory of shrubs and smaller trees, and a canopy of tall deciduous and evergreen trees. The disturbed forest edges were mended with species to match the surrounding areas. Of the hundreds of species he did incorporate, both native and nonnative, many did not survive due to unsuitable local conditions or lack of maintenance. Both the design plans and plant lists in each park are an invaluable reference for restoration efforts today.

Of all the parks, we know the most about the plants growing in Cherokee because of a floristic study conducted from 1936 to 1941. This study is of utmost importance historically because it has allowed us to see the changes in species composition over time. Little was known about the plants of the park until Mabel Slack published her master's thesis, "A Survey of the Flora of Cherokee Park at Louisville, Kentucky" (Slack 1941). Slack was an Atherton High School teacher and nature enthusiast who wanted to learn about the plants growing in her neighborhood park. Knowing very little about botany, she applied to graduate school at Cornell University and was accepted. Her advisor, Dr. Karl Wiegand, considered to be one of the greatest botanists of his time, approved her project in Kentucky. Mabel was granted permission by the Park Board to collect the woody and herbaceous plants of Cherokee, and she took to the field from

September 1936 through June 1941. At first, she collected "everything in sight as all was new." By the end, she had documented 523 taxa and collected over 1,100 herbarium specimens. Besides her own personal herbarium collection, a duplicate set was deposited at the Bailey Hortorum at Cornell University. This was the first published record of the plants of Cherokee Park.

In spring 2005, I began a study that entailed cataloging and collecting the plants of the park with a major emphasis on the herbaceous flora and woody invasive plants. With Slack's thesis as the backbone of my work, it was clear that changes had occurred over time. Wildflowers and ferns that were rare in 1941 and are no longer found today include fire pink, partridge-berry, starry campion, nodding ladies-tresses, maiden-hair fern, lily-of-the-valley, downy wood mint, and broad-leaved bluets. But surprisingly I cataloged a good number of plants that were rare in 1941 and are still rare today, including sharp-lobed hepatica, broad-leaved toothwort, American alum-root, purple-rocket, wild geranium, enchanter's nighshade, largeflower bellwort, green violet, purple cliffbreak, Eastern eulophus, and false garlic. A few of these are believed to be growing in the original locations discovered by Mabel Slack. Today, many native wildflowers, as well as trees and shrubs, are being replanted into the woodlands as part of the park's restoration efforts. The Louisville Chapter of Wild Ones is doing similar work in Wildflower Woods, the 7-acre parcel of land at the entrance of the park off Eastern Parkway.

Perhaps one of the most interesting plants ever found in Cherokee Park is running buffalo clover (*Trifolium stoloniferum*), which has a fascinating life history that is distinctly wedded to our heritage. This species is listed as threatened in Kentucky (KSNPC 2010) and is monitored by biologists from the Kentucky State Nature Preserves Commission in Frankfort. This member of the Legume Family looks similar to other weedy white clovers, but it has a pair of opposite leaves on the stem below the beautiful white flower heads, which bloom in early April to late May. Spreading by stolons, this clover is habitat specific, found only in areas underlain by limestone in Kentucky, Ohio, Indiana, West Virginia, and Missouri. It is associated with the old buffalo trails, traces, and prairies, where it depended on the bison for its success during the eighteenth and nineteenth centuries. These heavy creatures trampled the stolons, which would break into fragments and regenerate, forming new plants.

In 1990, a graduate student at the University of Kentucky discovered a small grouping of about four plants growing in thin turf near the basketball courts off Willow Avenue in Cherokee Park. It was sighted several other times from 1992 to 1997. However, after that year, it has never been seen again, despite efforts to monitor the location. This extremely rare plant needs filtered to partial sunlight, and the soil needs to be continuously disturbed, mimicking the days when the buffalo were plentiful and roamed the area. Today, it is most frequently observed in old cemeteries, lawns, and military lands in ten counties in Kentucky, all with very limited populations. It is hopeful that lightly tilling the soil might be a successful management practice to bring back this threatened Kentucky plant to Cherokee Park.

The major change in the flora appears to be the addition of nonnative plants since 1941. Some have exploded in their numbers just in the past eight years. Native to Europe and Asia, these troublesome plants are a major concern to the welfare of the woodlands at present. All are highly aggressive and successful in becoming established, despite routine effort, time, and money spent on measures to eradicate them. These invasive plants include amur honeysuckle, Asian bittersweet, Chinese privet, garlic penny-cress, lesser celandine, mugwort, Nepalese eualalia, porcelain-berry, Siberian squill, crabweed, field hedge parsley, and garlic-mustard. One species that was mentioned in 1941 but not problematic at the time was chocolate vine, or akebia. Today, this rampant ornamental vine smothers ground vegetation and climbs up and over the woodland vegetation forming tents, or shrouds, as tall as its support will allow. The movement and spread of these invasive plants are better documented in some cases than others, and more information is provided in the species accounts.

Maintaining the unique biodiversity of these parks, in conjunction with respecting Olmsted's planning philosophy and design, is at the core of the 2005 Woodlands Restoration Campaign. This campaign is led by our local Olmsted Parks Conservancy, whose ongoing mission is to "restore, enhance, and preserve" all the Olmsted Parks. Efforts to remove invasive plants, replant native species, address erosion issues from storm water run-off, and repair problematic trails are in place. However, these small urban green spaces are vulnerable and succumbing to the changing climate as repeatedly hot and dry summers and devastating winter wind and ice storms take a toll on the vegetation. A number of diseases such as those caused by the emerald ash borer and ambrosia beetle also threaten the vegetation, and these, along with everyday disturbance, are opening up gaps for the movement of new plant invasions like the Japanese chaff flower. This Asian herbaceous plant has just been found in Cherokee but has been running amok in thirteen other counties, especially along the Ohio River, where it is capable of establishing a monoculture in moist woods. The natural history of our parks' woodlands is rapidly changing, as witnessed in the eight years of this study.

Now, over a hundred years after the establishment of these parks, this field guide provides a snapshot of the native and transported plants in each one. It is my hope that this baseline information will be enlarged and amplified by future studies so that we might better understand how to "restore, enhance, and preserve" the natural heritage that makes up these magnificent, priceless Olmsted Parks, which have enriched the lives of the people of Louisville since the creation of our Olmsted Parks system in 1891.

ILLUSTRATED PLANT STRUCTURES

Parts of a dicot plant stem

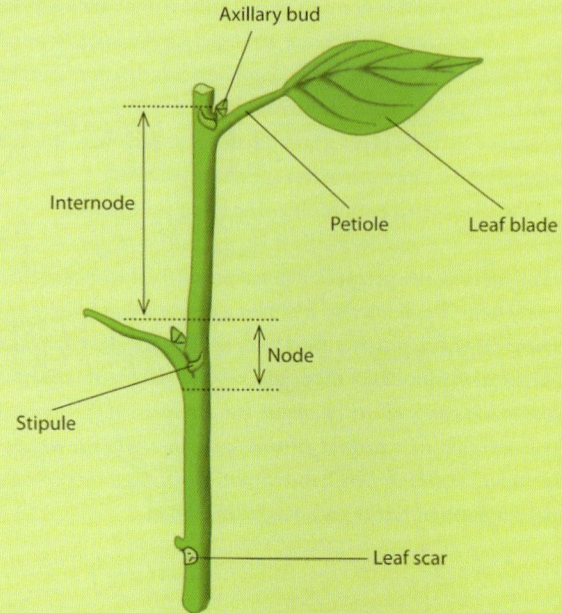

Axillary bud

Internode

Petiole

Leaf blade

Node

Stipule

Leaf scar

Parts of a monocot (grass) collar

*The collar region of grasses has the
most diagnostic vegetative characteristics.*

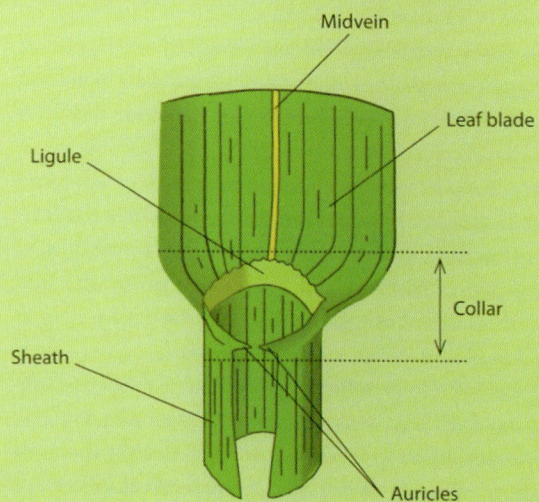

Midvein

Ligule

Leaf blade

Collar

Sheath

Auricles

Common Terminology

Using the family keys and understanding the species descriptions require a basic familiarity with plant structures and terminology. The most common characteristics and terms used to describe plants are given below. The glossary of botanical terminology at the back of the book includes definitions of terms used in the illustrations. Structures and growth habits important for identifying weeds and understanding the plant descriptions in this book are illustrated in this section.

Root types

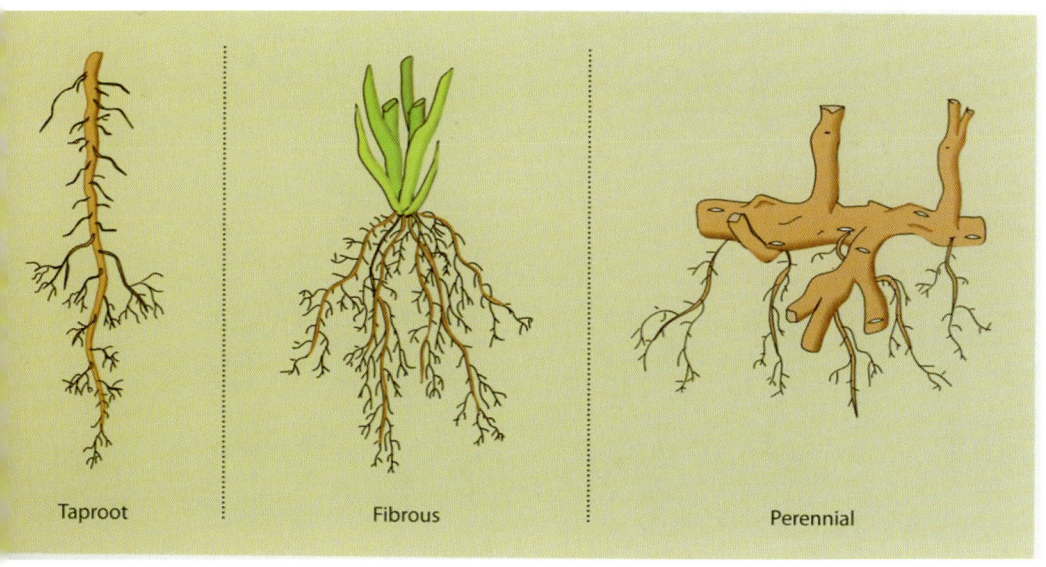

Taproot   Fibrous   Perennial

Perennial stem types

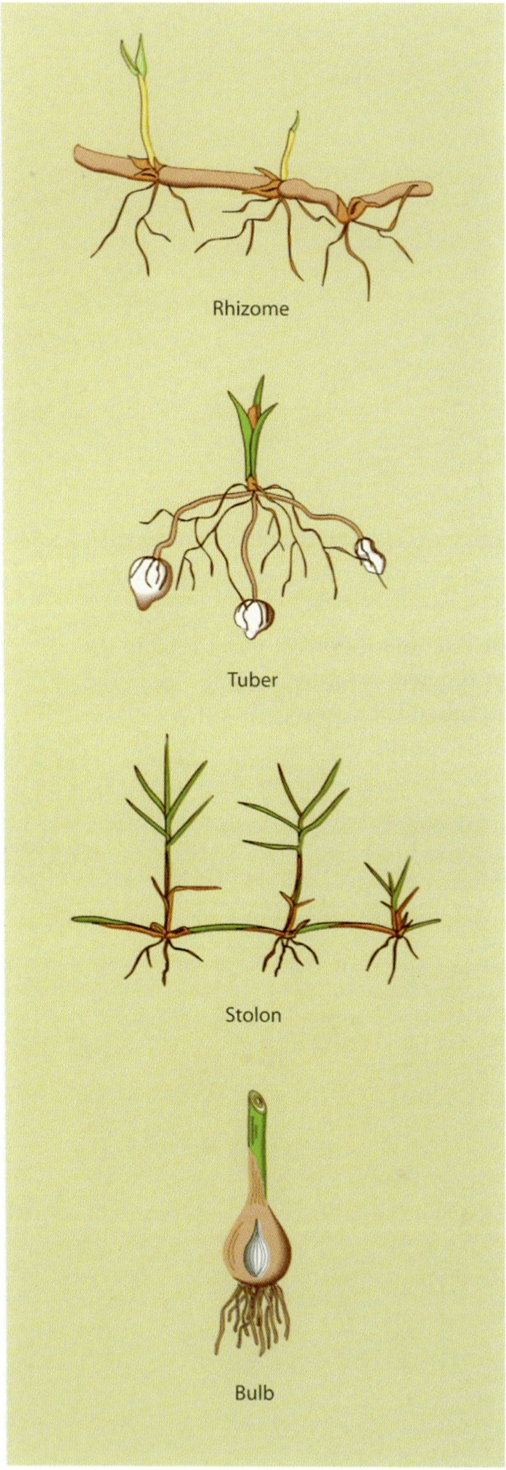

Rhizome

Tuber

Stolon

Bulb

Leaf arrangements

Alternate

Opposite

Basal rosette

Whorled

Leaf shapes

Elliptic

Lanceolate

Linear

Oblong

Obovate

Orbicular

Ovate

Reniform

Spatulate

Leaf margins

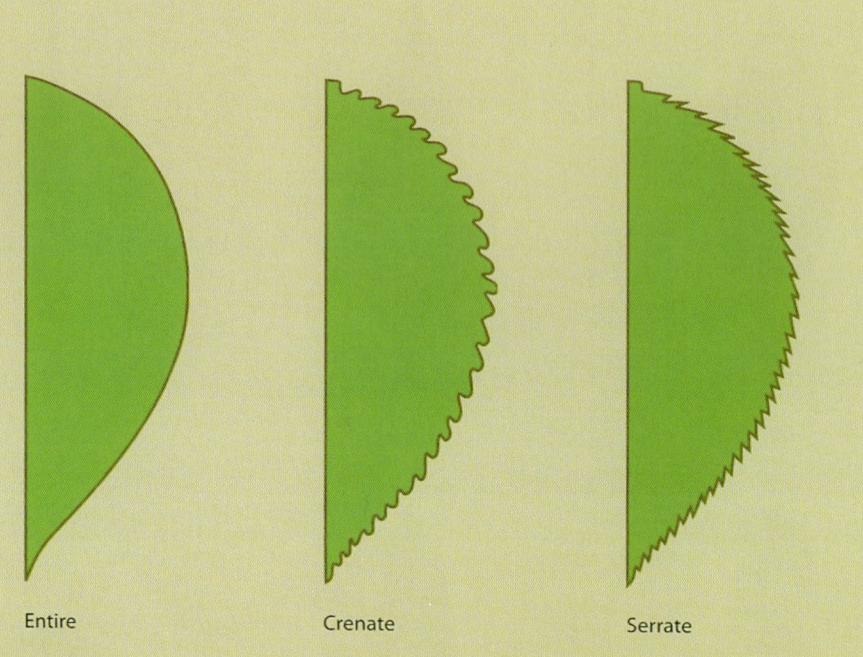

Entire   Crenate   Serrate

Palmate   Pinnate

Erect

Climbing, tendrils

Climbing, twining

Ascending

Decumbent

Prostrate

Parts of a dicot flower

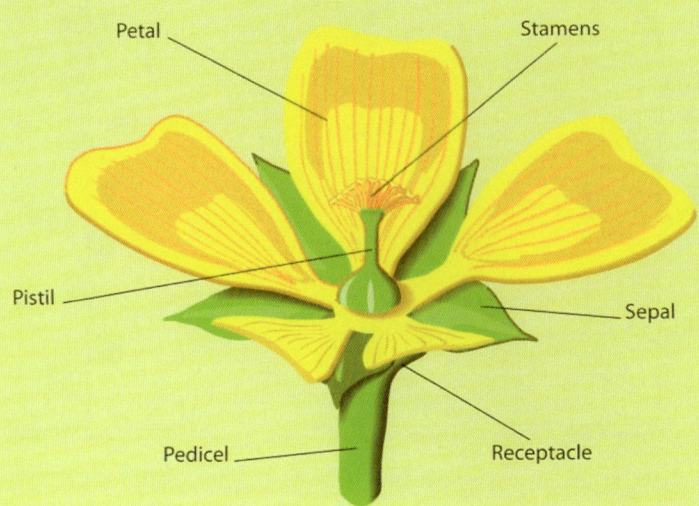

Petal

Stamens

Pistil

Sepal

Pedicel

Receptacle

Parts of a monocot floret

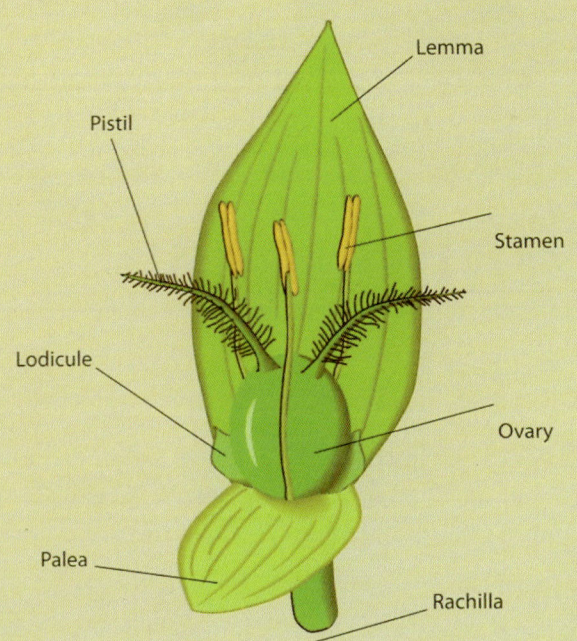

Lemma

Pistil

Stamen

Lodicule

Ovary

Palea

Rachilla

Inflorescences

Solitary

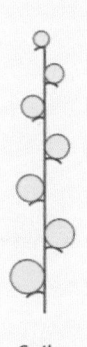

Spike

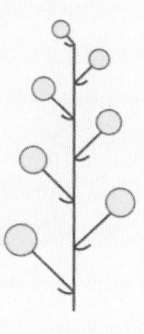

Raceme

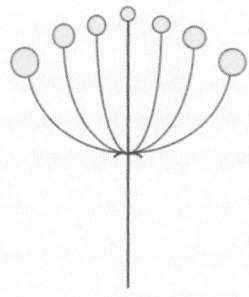

Umbel

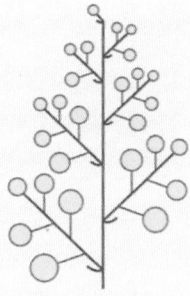

Panicle

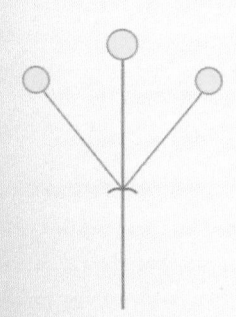

Simple cyme

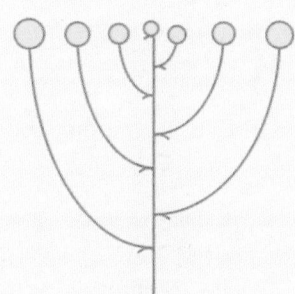

Corymb

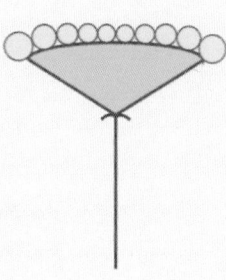

Head

PLANT
DESCRIPTIONS
of 384 Species

Ebony spleenwort
Spleenwort Family
Asplenium platyneuron (L.) BSP
Aspleniaceae

Key features: Fern from thick rhizomes; stems wiry, brittle, reddish brown; leaves with numerous paired leaflets.

Origin: Native.

Life form: Perennial fern.

Stems: Erect, unbranched, smooth, brittle for fertile blades, shorter and spreading for sterile fronds.

Leaves: Fertile ones dark green, erect, narrow, tapering to top and bottom, to 20 inches tall; sterile ones numerous, shorter, arched, lighter green; both with 15 to 22 pairs of leaflets.

Sporangia: In linear sori in 2 rows on underside of leaflets.

Distribution: Moist to dry woods, rocky ledges. Uncommon.

In Kentucky: AP, IP, ME.

This fern is found in southern Africa and in North America and is unique in that no other North American fern has such a distribution. In Kentucky, ebony spleenwort is one of the most widespread in the state and highly adaptable. It grows in a variety of habitats, from rich shaded woods to dry rocky masonry in urban sites. In old fields, it is often associated with Japanese honeysuckle and Eastern red-cedar.

The common name is misleading, as the smooth, dark stem is brown, not ebony black.

Walking fern

Spleenwort Family
Asplenium rhizophyllum L.
Aspleniaceae

Key features: Fern from scaly rhizome; stems flattened, in star-shaped tufts; blades triangular, tapering into a long, thin tip that roots upon soil contact.

Origin: Native.

Life form: Perennial fern.

Stems: Dark reddish brown at base, green and smooth above.

Leaves: Of 2 types: fertile ones to 15 inches long, often larger than the sterile ones; blade evergreen, shiny above, paler below, midvein pale, base heart-shaped to lobed.

Sporangia: In linear sori irregularly scattered along veins on underside of blade.

Distribution: Moist, mossy, shady limestone ledges. Rare.

In Kentucky: AP, IP, (ME-rare).

This unique fern is called walking fern because the blade with a long, pointed tip can sprout a new plant when it touches the ground and takes root. The old, slightly upright, arching fronds often are surrounded by many new young plants that grow flat to the ground.

The species name—*rhizo* is Latin for "root" and *phylum* Latin for "leaves"—refers to the way the plant spreads.

Southern bladder fern
Wood Fern Family
Cystopteris protrusa (Weath.) Blasdell
Dryopteridaceae

Key features: Fern from creeping rhizomes that protrude forward, rhizomes with golden yellow hairs; leaves in loose clumps, triangular to broadly lanceolate in general outline.

Origin: Native.

Life form: Perennial, from creeping rhizomes that extend forward before the emergent leaves.

Stems: Smooth, threadlike, straw-colored to green with small, tan scales near the fragile base.

Leaves: Erect to slightly arching, blades 3 to 10 inches long, leaflets small, 9 to 15, at right angles to stem or slightly ascending, opposite or nearly so, margins rounded to irregularly toothed, veins running to teeth not sinuses.

Sporangia: In round sori, few, scattered on lower surface.

Distribution: Open woods, limestone ledges. Mostly uncommon, but locally common in some areas of the park.

In Kentucky: AP, IP, ME.

This fragile-looking fern is common throughout the state. It produces two kinds of leaves: the first set in early spring is small and sterile, and the second set in late May or June is much larger and fertile. Spreading by rhizomes, it makes a verdant ground cover in a few spots at Cherokee.

Sensitive fern

Wood Fern Family
Onoclea sensibilis L.
Dryopteridaceae

Key features: Stems yellowish, winged above; sterile leaves with blades deeply lobed; fertile leaves are bare stalks with bead-like clusters of sporangia at the top.

Origin: Native.

Life form: Perennial from a stout, spreading rhizome.

Stems: Base brown, scales few, usually longer than blade; to 2 feet tall.

Leaves: Few, of 2 types: sterile leaves erect or ascending, deeply divided into 8 opposite, lobed pairs, light green, central stalk winged, blades net-veined, margins wavy, persisting all summer until frost; fertile leaves erect, with hard, bead-like clusters turning brown and persisting throughout winter.

Sporangia: In round sori hidden by inrolled leaflets.

Distribution: Wet woods along horse trail. Rare.

In Kentucky: AP, IP, ME.

Fossils of this fern dating back more than 60 million years show that the plants of today have changed very little in appearance over the millennia.

The common name refers to the sensitivity of the sterile leaves to frost.

Christmas fern

Wood Fern Family
Polystichum acrostichoides (Michx.)
Schott *Dryopteridaceae*

Key features: A fern from stout rhizomes; leaves dark green, arching, forming a circular cluster, leaflets with a distinct basal lobe or "toe."

Origin: Native.

Life form: Perennial fern.

Stems: Green, scaly.

Leaves: Several, to 2 feet tall; fertile leaves taller than sterile ones, upper portion bearing spores; each leaf cut into 20 to 40 lance-shaped, leathery leaflets with distinct basal lobe.

Sporangia: In round sori born in 2 rows on each side of midvein on underside of upper fertile leaves.

Distribution: Moist to dry wooded slopes, ravines. Uncommon.

In Kentucky: AP, IP, ME.

The fertile leaves of this fern are evergreen and a welcome site in the winter woods. The common name refers to the basal lobe or "toe" on the leaflets that resembles a stocking. The leaves are also gathered during the season and used for holiday decorations. The "fiddle heads" collected in the spring are used as food.

The Iroquois Indians made a poultice from the wet, smashed roots and applied them to the backs and heads of children with convulsions.

Common scouring-rush

Horsetail Family
Equisetum hyemale L. subsp. *affine*
(Engelm.) Calder & R.L.Taylor
Equisetaceae

Key features: Stems hollow, jointed, ridged, rough; leaves scalelike; sporangia in spore cases clustered in a terminal cone.

Origin: Native.

Life form: Perennial fern from branching rhizomes.

Stems: Erect, unbranched, sheaths at nodes with dark bands at top and bottom when mature, top band toothed, soon withering; fertile and sterile stems similar.

Leaves: Small, whorled.

Sporangia: Many, green, in a cone to ¹⁄₂ to 1 inch long, tip pointed, maturing in summer, splitting open to release the spores.

Distribution: Moist to wet ground. Uncommon.

In Kentucky: AP, IP, ME.

The rough stems, which feel like sandpaper, contain high amounts of silica and magnesium. These were used by the early settlers to give a satiny finish to wood and to scour metal or tin items, such as pots and pans—hence the name common scouring-rush.

This unique plant often forms large colonies, and the underground rhizomes grow very deep and are difficult to dig out. The plant is poisonous to livestock.

Common grape fern
Adder's-tongue Family
Botrychium virginianum (L.) Sw.
Ophioglossaceae

Key features: Fern often persisting into fall; leaf blade triangular in outline, horizontal, appearing lacy; sporangia in rounded bead-like cases.

Origin: Native.

Life form: Deciduous fern with fibrous roots.

Stems: Basal stalk erect, smooth, succulent, 4 to 8 inches long until meets the upper triangular lacy sterile blade; fertile stalk 6 to 20 inches tall above sterile blade.

Leaves: Sterile blade bright green, heavily dissected, becoming reflexed to almost horizontal; to 12 inches wide and long.

Sporangia: In spore cases clustered on ascending short branches at the top of a slender stalk rising above the sterile blade.

Distribution: Rich moist woods. Rare.

In Kentucky: AP, IP, ME.

Common throughout the state, this species is one of the first to appear in early spring and grows in dry to moist conditions. Although rare in both parks, the bright green, lacy sterile blade can be seen above the woodland floor, often-times persisting into fall. It is also called rattlesnake fern because the tip of the spore-bearing stalk resembles a rattlesnake tail when it first opens.

Southern adder's-tongue

Adder's-tongue Family
Ophioglossum vulgatum L.
Ophioglossaceae

Key features: Fern fleshy, withering in June; leaf a single, shiny blade, lily-like.

Origin: Native.

Life form: Perennial fern, withering in June, from thick, dark roots.

Stems: Single, upright, fleshy; to 12 inches tall.

Leaves: Single sterile blade erect to spreading, shiny, oval, widest below middle, net-veined, base tapering abruptly to slightly, tip rounded; 2 to 4 inches long appearing in April or May.

Sporangia: Round in 2 rows along margin of slender, pointed stalk.

Distribution: In low, moist woods southwest of stables. Rare.

In Kentucky: AP, IP, ME.

This unusual-looking plant is a thrill to find in low, moist woods at Iroquois. The single leaf blade, when first emerging in spring, looks like a lily or orchid leaf, but a closer look at the veins shows that they are not parallel but a complex network of veins.

The genus name—from the Greek *aphis* meaning "snake" and *glossa* meaning "tongue"—describes the slender, tonguelike shape of the spore-bearing portion of the plant.

Southern beech fern

Marsh Fern Family
Phegopteris hexagonoptera (Michx.)
Fee *Thelypteridaceae*
[*syn=Thelypteris hexagonoptera* (Michx.)
Weath]

Key features: A fern from black, scaly rhizomes; leaves broadly triangular in general outline with leaflet pairs opposite, joined at the base by a winged stem.

Origin: Native.

Life form: Perennial fern.

Stems: Slender, light green to reddish green or tan, smooth; to 20 inches tall.

Leaves: Both sterile and fertile leaves up to 13 inches long; leaflet pairs 12 to 24, opposite to subopposite each other; leaflets shallowly to deeply divided into 8 to 25 rounded lobes; lowest pair spreading outward and wider spaced from the one above.

Sporangia: In round sori scattered in a row on underside of leaflets near margin.

Distribution: Moist to wet woods, mostly east of stables. Uncommon.

In Kentucky: AP, IP, ME.

The genus name—from the Greek *phegos* "beech" and *pteris* "fern"—aptly describes the preferred habitat for this native fern, i.e., beech woods. Another common name is broad beech fern.

Spreading chervil
Carrot Family
Chaerophyllum procumbens (L.) Crantz
Apiaceae

Key features: Stems mostly smooth; leaves finely dissected; flowers tiny, white, in terminal and axillary umbels.

Origin: Native.

Life form: Annual herb from a slender taproot.

Stems: Ascending, weak, light green to brownish purple, branched from the base; to 2 feet tall.

Leaves: Lower ones stalked, with papery sheaths; three-times divided, leaf segments shallowly or deeply cleft into 3 to 5 linear to oblong lobes, tips rounded or short pointed; upper leaves reduced, clasping to short-stalked.

Flowers: Umbels with 3 to 7 flowers; sepals none; petals 5, white; stamens 5; ovary green. April through May.

Fruits: Composed of 2 dark brown, smooth, ribbed segments that are broadest at or near the middle, on threadlike stalks.

Distribution: Moist open woods, waterways, limestone ledges, thickets. Common.

In Kentucky: AP, IP, ME.

Similar species: Field hedge-parsley [*Torilis arvensis* (Huds.) Link] is an invasive annual plant from Europe that grows to 3 feet tall. The alternate leaves are **1 to 3 times pinnately divided with many coarsely toothed linear segments.** The small white flowers are produced in **umbels with 4 to 9 smaller umbellets.** Subtending the main umbel are either 1 to 3 very narrow linear bracts, or none. This species is found in all five parks but has spread rapidly in Cherokee, Seneca, and Iroquois in the past four years growing along roadsides, thickets, woodland edges, waterways, and fields. June through August. In Kentucky: IP. Listed as a Moderate Threat by the Kentucky Exotic Pest Plant Council.

Both species belong to an economically important family that is characterized by its distinct inflorescence, fruit, odor, flavor, and toxicity. Although spreading chervil and field hedge-parsley are not used in cooking or for flavoring, they are related to anise, cumin, dill, fennel, and coriander.

Harbinger-of-spring

Carrot Family
Erigenia bulbosa (Michx.) Nutt.
Apiaceae

Key features: Plant from an underground tuber; leaves basal and alternate, divided; flowers tiny, white with reddish pink anthers turning black.

Origin: Native.

Life form: Perennial herb.

Stems: Erect to ascending, one or more from the base, slender, purplish, smooth; 3 to 10 inches tall.

Leaves: Leaflets 3, each with many narrow, irregularly cleft to 3-lobed segments, dull green above, shiny below, margins entire, tips pointed; 3 to 8 inches long.

Flowers: In clusters, with 1 to 6 flowers; sepals absent; petals 5, narrow; stamens 5; style white, divided; ¼ inch wide. Late January through March.

Fruits: Tiny, laterally flattened, incurved at top and bottom, 5-ribbed.

Distribution: Rich, moist woods. Uncommon.

In Kentucky: AP, IP, (ME-rare).

This delicate, early-blooming wildflower is a true harbinger of spring and is barely visible above the withered brown leaf litter of the previous autumn. Also known as salt and pepper, the reddish pink anthers turn black when dried and stand out against the small white petals.

Other common names are ground-nut and turkey-pea and refer to the small underground tuber that can be eaten raw or cooked. *Erigenia* means "born in the spring."

Plantain pussytoes
Aster Family
Antennaria plantaginifolia (L.) Richardson
Asteraceae

Key features: Plant from stolons, often forming colonies; leaves densely white woolly below; flowering stalks and flowering heads white woolly.

Origin: Native.

Life form: Perennial herb.

Stems: Lacking except for the flowering stems 3 to 12 inches tall.

Leaves: Blades obovate to lanceolate, margins entire, veins 3 to 5, dull green above, woolly below; alternate stem leaves remote, narrow, clasping, reduced upwards; 1 to 3 inches long.

Flowers: Heads several, terminal, 1/4 to 1/2 inch wide with male and female flowers on different plants; disk florets tubular, with either 5 stamens and purplish brown anthers or female with a 2-forked, purple-tipped style; involucral bracts in 1 to 2 series, covered in cobwebby hairs. April through June.

Fruits: Nutlets brown, small, with sticky dots; hairs white, tufted.

Distribution: Dry open woods, woodland edges, roadsides in thin turf. Rare at Cherokee; common at Iroquois.

In Kentucky: AP, IP, ME.

The name pussytoes refers to the white, woolly female flower clusters that look soft like a kitten's paw, while the name dog's toes refers to the male flowers. Another common name is women's tobacco and was used by the Native Americans as a gynecological aid. An infusion was given to help prevent sickness after childbirth. The roots and tops were taken for colds and coughs. Country folks were said to pack the dried flower heads with their stored woolens to keep the moths out. It was also used in shampoo to banish lice.

Dense colonies can be seen growing in poor soil and spreading by stolons from the mother plant; contains allelopathic chemicals that inhibit the growth of other plants nearby.

Philadelphia fleabane

Aster Family
Erigeron philadelphicus L.
Asteraceae

Key features: Stems leafy; leaves alternate, clasping above; flower heads terminal, nodding in bud, outer ray florets white or pink, 100 to 150.

Origin: Native.

Life form: Biennial or short-lived perennial from a fibrous taproot.

Stems: Erect, slightly ridged, mid-stem hairs long, spreading; 1 to 3 feet tall.

Leaves: Basal, blades oblanceolate to obovate, to 6 inches long, margins sharp to bluntly toothed or lobed, tapering into a stalk; stem leaves reduced upward.

Flowers: Heads 1 to many, about 1 inch wide; inner disk florets yellow; outer ray florets very narrow, straplike; involucral bracts green, narrow, margins papery. April through July.

Fruits: Achenes tiny, tan to brown, 2-nerved with bristles.

Distribution: Disturbed ground, fields, woodland edges, thickets, roadsides. Common.

In Kentucky: AP, IP, ME.

Similar species: Annual fleabane [*Erigeron annuus* (L.) Pers.], a native winter annual to 3 feet tall and with **narrow leaves that do not clasp the stem.** Flower heads are terminal and **do not nod in bud.** There are many with yellow inner disk florets and **50 to 100 white to purplish pink outer ray florets.** Cher, Sen, Iroq, Shaw, Chick. Disturbed ground in fields, woodland edges, thickets, roadsides. Common. June through September. In Kentucky: AP, IP, ME.

The generic name is derived from the Greek words *eri* "early" and *geron* "old man"—or "old man in the spring"—and refers to the gray appearance of some of the species. Both species above can cause contact dermatitis.

Twin-leaf

Barberry Family
Jeffersonia diphylla (L.) Pers.
Berberidaceae

Key features: Leaves long-stalked, bluish gray, divided into two unequal kidney-shaped divisions; flowers solitary, white; fruit capsules opening like a lid at top.

Origin: Native.

Life form: Perennial from a rhizome and fibrous roots.

Stems: Lacking except for the flower stalks to 12 inches tall.

Leaves: Upright, few, basal, long-stalked; leaflets each about 1½ inches wide, enlarging to 3 to 6 inches wide after flowering.

Flowers: Terminal on smooth leafless stalks extending above leaves; sepals 4, falling early; petals 8, narrowed at base; stamens 8; pistil 1; 1 to 2 inches wide. March through early April.

Fruits: Capsule obovoid, opening by a horizontal split; seeds many, in rows.

Distribution: Moist wooded slopes, especially in limestone soils. Uncommon.

In Kentucky: (AP-rare), IP.

The generic name honors Thomas Jefferson, naturalist and third president of the United States. The species name combines the Greek word *di* "two" and the Latin word *phyllon* "leaf," and this describes the leaf that is deeply divided as if to have two blades.

Native Americans used the roots in a wash that was used externally for joint pain, sores, cancerous sores, ulcers, and inflammation.

May-apple

Barberry Family
Podophyllum peltatum L.
Berberidaceae

Key features: Leaves simple or with 5 to 9 deeply cut lobes; flowers solitary, white, waxy, nodding from fork of leaves; fruits yellow, lemon-like.

Origin: Native.

Life form: Perennial herb from a rhizome often forming large colonies.

Stems: Lacking except for the flowering stalk 12 to 18 inches tall.

Leaves: Peltate, veins pale, margins toothed; 6 to 15 inches wide; young plants with one leaf, flowering plants with two leaves.

Flowers: Sepals 6, falling early; petals 6 to 9; stamens 12 to 18, yellow; ovary ovoid, stigma large, thick; 1 to 2 inches wide. April through May.

Fruits: Berry pulpy, to 2 inches long; seeds many.

Distribution: Moist woods, woodland edges, waterways, thickets, ravine slopes. Common at Cherokee; uncommon elsewhere.

In Kentucky: AP, IP.

Often called umbrella leaf, the young plants, which are visible in spring, are tightly wrapped around the stem in a manner of a closed umbrella. The genus *Podophyllum* means "foot leaf" and refers to the stout stem, while *peltatum* describes the "shield-shaped" leaves.

The Native Americans would place a drop of juice from the fresh root in the ear to heal deafness. The Iroquois Indians would mix the leaves with other medicinal plants and soak the corn seeds in this before planting. A similar practice was used by the Cherokee Indians—they would soak the corn seeds in "root ooze" before planting. This practice was believed to ward off crows and insects.

Although the ripe fruit pulp is edible, the seeds and rhizomes are poisonous.

Garlic-mustard

Mustard Family
Alliaria petiolata (M.Bieb.) Cavara &
Grande
Brassicaceae

Key features: Plant smells of garlic;
leaves basal and alternate, distinctly
veined; flowers white, small in terminal
and axillary clusters.

Origin: Europe.

Life form: Biennial herb from a slender
taproot.

Stems: Erect, single or little-branched,
light green, smooth above, hairy below;
1 to 4 feet tall.

Leaves: Basal rosette of heart-shaped,
bluntly toothed leaves, 1 to 3 inches wide
and long; stem leaves alternate, blades
heart-shaped to triangular, margins
coarsely to bluntly toothed, all on slender
stalks; 1½ to 3 inches long.

Flowers: Sepals 4, green; petals 4,
narrowed at the base; stamens 6; pistil 1.
April through July.

Fruits: Slender, erect or ascending,
4-angled, 1 to 3 inches long; seeds tiny,
black, arranged in a row.

Distribution: Dry to moist woods,
limestone ledges, fields, waterways, along
roadsides. This plant is especially invasive
in moist woods at Cherokee, Seneca, and
Shawnee.

In Kentucky: (AP-rare), (IP-mostly
Bluegrass). Listed as a Severe Threat by the
Kentucky Exotic Pest Plant Council.

Invasive plant
Cher, Sen, Iroq, Shaw, Chick

This European native was first discovered in the United States in 1869 on Long Island, New York. It was introduced as a pot herb because the leaves are rich in Vitamin A and C and can be boiled in soups or used in salads.

The seedpods turn tan by midsummer and split open, forcefully expelling the seeds as far as ten feet from the parent plant. Especially troublesome in moist woods, newly established plants colonize rapidly and compete with the native spring wildflowers for light, nutrients, and space. In Kentucky, this species has increased rapidly in the Bluegrass Region since early records in the 1970s.

In England, it is called Jacke of the hedges or poor-man's mustard because the poor country folks would eat the leaves with bread or mix them with lettuce as well as make a stuffing for pork.

The common name refers to the odor of garlic when crushed.

Smooth sickle-pod

Mustard Family
Arabis laevigata (Muhl.) Poir.
Brassicaceae

Key features: Stem leaves alternate, leaf bases arrow-shaped, clasping the stem; flowers small, white; fruit pods elongating, often curving downward.

Origin: Native.

Life form: Biennial herb from a taproot.

Stems: Erect, single or branched, smooth with a powdery, grayish green bloom; 1 to 3 feet tall.

Leaves: Basal rosette in first year; blades ovate, sharply to bluntly toothed or entire; stem leaves narrow, margins toothed, up to 8 inches long; smaller upward.

Flowers: Stalked, in loose terminal and axillary clusters, elongating in fruit; sepals 4, light green; petals 4, equaling or slightly longer than sepals; stamens 6, anthers yellow; pistil 1; ¼ inch wide. April through June.

Fruits: Narrow spreading horizontally to arching downward, to 4 inches long; seeds tiny, winged, in a row.

Distribution: Rocky woods, moist woods, limestone ledges. Common.

In Kentucky: AP, IP, ME.

The common name refers to the distinctive seedpods that droop downward resembling the blade of a sickle.

Shepherd's purse
Mustard Family
Capsella bursa-pastoris (L.) Medik
Brassicaceae

Key features: Leaves mostly basal, divided or lobed; flowers white, tiny; seedpods flattened, triangular or heart-shaped.

Origin: Europe.

Life form: Winter annual from a taproot.

Stems: Erect, sparingly branched, rough-hairy; to 24 inches tall.

Leaves: Variable; the lower ones in a basal rosette to 9 inches long, segments deeply cleft to slightly lobed; alternate stem leaves few, narrow, arrow-shaped, clasping; 1 to 2 inches long.

Flowers: In terminal clusters on slender stalks; sepals 4; petals 4; stamens 4; pistil 1. March until frost.

Fruits: Seedpods on threadlike stalks to 1 inch long, horizontal to mostly ascending; seeds oblong, shiny orange-brown.

Distribution: Disturbed ground in moist open woods, roadsides, fields, turf, cultivated beds. Common weed.

In Kentucky: AP, IP, ME.

The common name refers to the resemblance of the flat seedpods to an old-fashioned change purse. The dried leaves and fruit pods, slightly peppery in taste, are used for seasoning soups and salads.

In the Middle Ages, shepherd's purse was a "blood" herb. It was said to stop the flow of blood anywhere on the body, even if the plant was only held in the hand. Today, herbalists use it for the supposed ability to heal open wounds and hemorrhages. It is also known traditionally for its uterine-contracting properties and use in childbirth.

Hoary bitter-cress

Mustard Family
Cardamine hirsuta L.
Brassicaceae

Key features: Leaves mostly basal and lobed, stalks hairy at base; flowers white, tiny; seedpods slender, ascending.

Origin: Eurasia.

Life form: Winter annual from a slender taproot.

Stems: Several, slender, light green, grooved, short when young, elongated to 11 inches tall.

Leaves: Basal rosette forms a circle 2 to 5 inches wide; lower stem leaves with 1 to 3 pairs of rounded, slightly lobed leaflets, terminal lobe largest; alternate stem leaves, smaller with narrow segments.

Flowers: In terminal clusters; sepals 4, light green to purplish; petals 4, very narrow; stamens 4; pistil 1. February through May.

Fruits: Very narrow to $1\frac{1}{2}$ inches long, often ascending past the flowers.

Distribution: Disturbed ground, especially in turf, moist woods, waterways, fields, cultivated beds. Common weed.

In Kentucky: AP, IP, ME.

Also called hairy bitter-cress. The common name refers to the hairs at the base of the leaf stalk, which aid in the identification of this weedy plant. The persistent, lacy basal leaves are easy to spot, especially when many plants are clustered together and form a dense mat in moist shady woods and in sunny disturbed ground. It is an agricultural weed in pastures and small grains.

The generic name is derived from the Greek word *kardamon,* which refers to a plant in the Mustard Family.

Five-parted toothwort

Mustard Family
Dentaria laciniata Muhl. ex Willd.
Brassicaceae

Key features: Plant from jointed, knobby rhizomes with fleshy tubers; stem leaves in a whorl of 3, divided into 3 to 5 narrow lobes; flowers white, pink, or lavender.

Origin: Native.

Life form: Perennial herb.

Stems: Erect, slender, hairy above; to 15 inches tall.

Leaves: Basal similar to stem leaves but withering early; upper ones deeply divided into 3 to 5 linear-lanceolate lobes, margins irregularly toothed; to 2½ inches long.

Flowers: Few, in terminal clusters; sepals 4, green to purple; petals 4; stamens 4, anthers yellow; pistil 1; ¾ inch long. March through May.

Fruits: Narrow, ascending, long-beaked; seeds tiny, in a single row.

Distribution: Moist woods. Common in Cherokee and Iroquois: rare in Seneca and Shawnee.

In Kentucky: AP, IP, ME.

Similar species: Broad-leaved toothwort (*Dentaria diphylla* Michx.) is a perennial herb to 18 inches tall with **long slender underground rhizomes.** There is a **pair of opposite or nearly opposite leaves with 3 elliptic blades** with bluntly toothed segments. Flowers are white to pink, small, in loose terminal clusters. Cher. Moist woods, mossy limestone ledges. Rare. March through May. In Kentucky: AP, IP.

Broad-leaved toothwort is also called pepper root (although it tastes like horseradish) or toothache root. The roots, which have toothlike swellings on them, were used in folk medicine as a cure for toothaches.

The Native Americans were said to have chewed the roots to cure a cold.

Whitlow-grass

Mustard Family
Draba verna L.
Brassicaceae

Key features: Plant delicate; leaves basal, entire; flowers tiny, white; fruit pods small, on threadlike stalks.

Origin: Europe.

Life form: Winter annual from fibrous roots.

Stems: One to several, threadlike, erect to ascending; to 8 inches tall.

Leaves: In a basal rosette; blades oblanceolate or spatulate, green to purple-tinged, densely hairy, margins entire, tips rounded to short pointed; ½ to 1½ inches long.

Flowers: In clusters at the tips of the stems; sepals 4, green, ovate; petals 4, white, deeply cleft; stamens 4. February through April.

Fruits: Pods oblong to elliptic, veined, less than ½ inch long, on threadlike stalks; seeds orange-brown, as many as 60 per pod.

Distribution: Disturbed ground in fields, turf, roadsides, cultivated beds. Common weed.

In Kentucky: (AP-rare), IP, (ME-rare).

Similar species: Short-fruited whitlow-grass (*Draba brachycarpa* Nutt. ex Torr. & A. Gray) is a native winter annual to 6 inches tall. It has **basal and stem leaves** that are **entire margined to slightly toothed**. The small white flowers are produced in a raceme that elongates with maturity. The fruit stalks are shorter than or equaling the tiny fruit. Cher, Sen. In thin soil over limestone, fields. Rare. February to April. In Kentucky: (AP-rare), IP, ME.

Both of these species are easily overlooked because of their small size. However, whitlow-grass is common in pastures, especially in central Kentucky.

It is a true harbinger of spring, sometimes flowering in late January. The species name, *verna,* means "spring."

This European plant has been used to treat Whitlow disease, an inflamed sore on the toe, fingernail, or animal hoof— hence the common name whitlow-grass. Other common names are nailwort, shad-blossom, and white blow.

Short-fruited whitlow-grass is rare in Kentucky but seems to be spreading via contaminated top soil.

Garlic penny-cress
Mustard Family
Thlaspi alliaceum L.
Brassicaceae

Key features: Plant smells of garlic; leaves basal below, alternate above, bases arrow-shaped, clasping; flowers tiny, white, in terminal and axillary clusters.

Origin: Europe.

Life form: Summer annual herb from a taproot.

Stems: Erect, single or multiple from base, smooth above, basal hairs long; to 18 inches tall.

Leaves: Basal with terminal lobe largest; stem leaves alternate; blades oblong,

margins irregularly toothed, smooth on upper surface; 1 to 2 inches long.

Flowers: Clusters compact at first, soon elongating; sepals 4, green, shorter than petals; petals 4, erect to spreading; stamens 6, anthers yellow; pistil 1, green; flower stalks elongating in fruit to ¾ inch long. April through June.

Fruits: Pods ovate, 2-valved, green, smooth, narrowly winged, tip notched; seeds 2 to 4, tiny, dark brown when mature.

Distribution: Roadsides, waterways, floodplains, wet fields, cultivated beds.

In Kentucky: Collected in 6 counties, including Jefferson, and 1 in Indiana. Listed as a Significant Threat by the Kentucky Exotic Pest Plant Council.

This weedy species was first found in North America in the 1960s when it was collected in North Carolina. Since then, it has spread through much of the eastern United States and south to Louisiana. In Kentucky, it was not known to be a part of the flora until 1985. The genus has two other nonnative species that are widespread in the United States. One in particular, perfoliate penny-cress (*Thlaspi perfoliatum*), looks very similar to garlic penny-cress. The presence of garlic penny-cress in Kentucky was first noted by John Thieret in 1985, from specimens he collected in 1982 and 1984 but were misidentified.

Also called roadside penny-cress, this plant often grows with the other two

Invasive plant
Cher, Sen, Iroq, Shaw, Chick

penny-cresses along roadsides, in pastures, fallow fields, waterways, and other disturbed sites.

In Cherokee, it has spread rapidly along Beargrass Creek since it was first documented in 2006 and has become an aggressive plant. It is also spreading rapidly along the banks of the Ohio River.

Japanese honeysuckle

Honeysuckle Family
Lonicera japonica Thunb.
Caprifoliaceae

Key features: A vine; leaves opposite, deeply pinnately lobed when young, entire when mature; flowers white or pink, fading to yellow with age.

Origin: Asia.

Life form: High-climbing, twining, or trailing vine from underground rhizomes and runners.

Stems: Young stems green, becoming reddish brown with age, pith hollow; to 20 feet or more.

Leaves: Short-stalked, blades ovate to oblong, margins entire to slightly lobed; to 3 inches long.

Flowers: One or 2 in leaf axils; petals tubular, 2-lipped, upper lip 4-lobed, lower lip 1-lobed; stamens 5; pistil 1, projected outward and curved; a pair of leaflike bracts subtend the flowers; 1 to 2 inches long. April through July.

Fruits: Berry black, round, ¼ inch wide; seeds 2 to 3, flattened.

Distribution: Disturbed ground, especially in open woods, woodland edges, fields, thickets, cultivated beds, roadsides.

In Kentucky: AP, IP, ME. Listed as a Severe Threat by the Kentucky Exotic Pest Plant Council.

Japanese honeysuckle was introduced into the United States in 1806 as a garden ornamental. It was also used for erosion control and wildlife cover. The leaves are eaten by rabbits and deer, and birds relish the fruits. It has now become naturalized and spread throughout much of the United States.

Although introduced from Japan, some authorities believe it is originally native to China. In traditional Chinese medicine it was also used to treat colds and flu.

Common mouse-ear chickweed
Pink Family
Cerastium vulgatum L.
Carophyllaceae
(*syn=Cerastium fontanum* Baumg.)

Key features: Plant mat-forming; stems sticky-hairy; leaves opposite, entire; flowers small, white, with 5 deeply notched petals.

Origin: Eurasia.

Life form: Perennial herb from fibrous roots.

Stems: Erect, ascending or sprawling, often rooting at the lower nodes, slender; to 20 inches tall.

Leaves: Sessile, in pairs of 3 to 7, blades oblong to ovate-lanceolate, 1-nerved, hairy, tips blunt; ½ to 1 inch long.

Flowers: Two to several, in terminal clusters; sepals 5, hairy, margins membranous, tips pointed, lacking long hairs; petals 5, sometimes absent; stamens 10; pistil 1. April through October and sporadically until November.

Fruits: Capsules very small, cylindrical, curved; seeds many, minute, reddish brown, knobby.

Distribution: Disturbed ground, especially in turf, fields, roadsides, cultivated beds. Common weed.

In Kentucky: AP, IP, ME.

This weedy species was introduced from Eurasia and is widespread throughout most of the United States, where it is a troublesome weed in crops, cultivated beds, and turf. The tiny seeds are long lived and may remain dormant in the soil for several years.

The generic name comes from the Greek word *keras* "horn" and refers to the hornlike seed capsules.

Common chickweed
Pink Family
Stellaria media (L.) Vill.
Caryophyllaceae

Key features: Stems often mat-forming and rooting at the nodes; leaves opposite, entire; flowers white, petals deeply notched, solitary or in small clusters.

Origin: Eurasia.

Life form: Winter annual herb from fibrous roots.

Stems: Slender, erect, hairs in vertical lines; to 20 inches tall.

Leaves: Blades ovate to elliptic, tips sharp pointed; lower ones similar, short-stalked, hairy at base; ½ to 1 inch long.

Flowers: Sepals 5, green, long; petals 5, short; stamens 3 to 10; pistil 1, styles 3 to 4; to ½ inch wide. March through November, capable of flowering all year round.

Fruits: Capsules small, ovoid; seeds dull reddish brown with curved rows of bumps.

Distribution: Disturbed ground in damp woods, fields, roadside ditches, turf, cultivated beds. Common to locally abundant in moist ground.

In Kentucky: AP, IP, ME. Listed as a Severe Threat by the Kentucky Exotic Pest Plant Council.

An old wives' remedy said to be effective against obesity was to drink common chickweed water or tea. It was also used for troubled stomachs and as a remedy to allay itching and burns when applied externally.

This cosmopolitan weed is found throughout the world. It is rich in copper, phosphorous, and iron and is eaten as a salad green or vegetable in many countries and tastes like spring spinach.

Also known as chicken-weed, Indian-chickweed, and white-birdseye, it is a heavy seed producer, and the whole plant is relished by wild birds and domestic fowl for food.

Great chickweed
Pink Family
Stellaria pubera Michx.
Caryophyllaceae

Key features: Plant with fertile and infertile shoots; leaves opposite, narrow; flowers white, petals deeply cleft, in loose clusters.

Origin: Native.

Life form: Perennial herb from fibrous roots and stolons.

Stems: Slender, erect, ascending or reclining, green to purplish green, occasionally branched at tips; 5 to 15 inches tall.

Leaves: Blades ovate-elliptic or lanceolate, margins entire, bases tapering or short-stalked, tips short pointed; 1 to 3 inches long.

Flowers: Sepals 5, green, tips long-pointed; petals 5; stamens 10, anthers brown; styles 3; ½ inch wide. April through June.

Fruits: Capsules ovoid; seeds tiny, yellowish brown, kidney-shaped.

Distribution: Moist woods, ravine slopes, ledges. Common in Cherokee; uncommon at Iroquois.

In Kentucky: AP, IP, (ME-rare).

This beautiful wildflower produces both fertile and infertile shoots. In mid-summer, a vigorous, large-leaved shoot arises and does not produce flowers and is different in form from those in spring: these sterile shoots can make identification difficult at first glance.

The generic name derives from the Latin *stella* "star" and refers to the shape of the white flower.

False solomon's-seal

Lily-of-the-Valley Family
Maianthemum racemosum (L.) Link
Convallariaceae
[*syn=Smilacina racemosa* (L.) Desf.]

Key features: Stems zigzagged between leavés; leaves spreading horizontally in two ranks; flowers small, white, in dense terminal clusters to 4 inches long.

Origin: Native.

Life form: Perennial herb from fibrous roots and stout rhizomes.

Stems: Simple, arching, light green; 1 to 3 feet tall.

Leaves: Alternate, clasping to short-stalked; blades elliptic-oblong, veins 3 to 5, distinct, margins entire, abruptly pointed at tip; 3 to 6 inches long.

Flowers: Tepals 6, white, spreading; stamens 6, anthers pale yellow; ovary white, style short. April through June.

Fruits: Berries purple spotted, turning dark red at maturity, ¼ inch wide.

Distribution: Moist woods, slopes. Uncommon at Iroquois; common at Cherokee.

In Kentucky: AP, IP, ME.

This graceful spring wildflower is called false solomon's-seal because it is said that it doesn't have the esteemed medical or magical powers of the true or smooth Solomon's-seal (*Polygo natum biflorum*). Other common names are starry Solomon plume and beadrubbies.

An infusion made from the leaves has been used in folk medicine as a means of birth control. The Cherokee Indians used the roots to make a cold steep to bathe sore eyes and the smoke from the roots calmed hysterics.

Smooth solomon's-seal

Lily-of-the-Valley Family
Polygonatum biflorum (Walter) Elliott
Convallariaceae
[*syn=Polygonatum commutatum*
(Schult. f.) A. Dietr.]

Key features: Stems arching to one side; leaves alternate, 2-ranked; flowers whitish green, drooping below the leaves on slender stalks.

Origin: Native.

Life form: Perennial herb from a knotty rhizome.

Stems: Solitary, leafy above, smooth; 1 to 3 feet tall.

Leaves: Blades lance-elliptic, sessile, green above, grayish blue and smooth below, veins parallel, margins entire; to 5 inches long.

Flowers: One to 5, in axillary clusters; tepals whitish green, tubular, tips 6-lobed, small, recurved; about $1/2$ inch long. April through June.

Fruits: Berry black or blue; seeds several.

Distribution: Moist woods, ravine slopes. Common in Cherokee; uncommon in Iroquois and Seneca.

In Kentucky: AP, IP, ME.

The stout rhizome was made into a tea and used externally as a wash for pains, cuts, sores, and bruises. It is an anti-inflammatory as well as an astringent, and it is said to heal any wound in a day or two. It also contains saponins, which are glycosides with a distinctive foaming ability. Because of the soapy properties, the plant has been used in shampoos, facial cleansers, and cosmetics.

Three-leaved sedum

Stonecrop Family
Sedum ternatum Michx.
Crassulaceae

Key features: Low-creeping, succulent woodland herb; leaves mostly in whorls of 3; flowers small, white, crowded on one side of a 2-to-4-branched terminal cluster.

Origin: Native.

Life form: Perennial herb from rhizomes and vegetative offshoots.

Stems: Light green, creeping with a single flowering stem and several leafy, short, infertile shoots; 4 to 10 inches long.

Leaves: Alternate above, blades oblanceolate to oblong, smaller than the flowerless terminal rosette of 6 crowded leaves with blades round to obovate.

Flowers: Sepals 4 to 5, narrow, bases united; petals 4 to 5, white, separate, twice as long as sepals; stamens 8 to 10, anthers dark red; pistils distinct. April through June.

Fruits: Follicle; seeds numerous.

Distribution: Damp limestone ledges, mossy banks. Rare.

In Kentucky: AP, IP, ME.

Similar species: Golden-carpet (*Sedum acre* L.) is a mat-forming succulent groundcover that has escaped cultivation and is becoming established **along roadsides and rocky banks** in Cherokee. From Eurasia, the many **small, bright yellow flowers**, ½ inch wide, form a colorful mass in May through July against the many **alternate, lime green leaves.** Rare, but is spreading. In Kentucky: Only collected from 3 counties: Jefferson, Oldam, McCreary.

The genus name, *Sedum,* is from *sedere,* meaning "to sit," and describes the manner in which many species attach themselves to rocks and walls.

Three-leaved sedum is also called woodland stonecrop and wild stonecrop.

Squirrel-corn
Fumitory Family
Dicentra canadensis (Goldie) Walp.
Fumariaceae

Key features: Plant poisonous, from yellow pea-shaped underground bulblets; leaves finely dissected; flowers white, spurs heart-shaped.

Origin: Native.

Life form: Perennial herb.

Stems: Lacking except for the flower stalk to 12 inches long.

Leaves: Lacy, segments narrow, finger-like, bluish green often with a whitish cast; on slender stalks either upright or slightly arching; to 10 inches long.

Flowers: Three to 12 hanging above the leaves, each ½ inch long; sepals 2, scalelike, falling early; petals 4, in 2 series, the outer 2 forming a nectar spur, inner 2 petals narrow, crested with tips united over the stigma; stamens 6. March through April.

Fruits: Capsule oblong to linear; seeds 10 to 20, crested.

Distribution: Moist woods. Common in Cherokee growing with Dutchman's-breeches; rare in Iroquois.

In Kentucky: AP, IP.

This fragrant spring wildflower is also called Indian-potatoes and turkey corn and describes the small yellow bulblets that resemble grains of corn or peas. These bulblets are poisonous to cows and horses that graze in wooded pastures in early spring and eat them.

Another common name is ghost corn because some Native Americans believed it was food for spirits.

Dutchman's-breeches

Fumitory Family
Dicentra cucullaria (L.) Bernh.
Fumariaceae

Key features: Plant poisonous, from pink or white underground bulblets; leaves basal, lacy; flowers white, spurred with bases bluntly pointed and spreading.

Origin: Native.

Life form: Perennial herb.

Stems: Lacking except for the flowering stalk 5 to 12 inches long.

Leaves: Dissected into narrow fingerlike segments, gray-green above, silvery green below; stalks upright to slightly arching; to 10 inches long.

Flowers: Several, in a raceme dangling above the leaves; sepals 2, tiny; petals 4, outer 2 white; inner petals 2, pale yellow, crested; stamens 6; $1/2$ to $3/4$ inch long. March through May.

Fruits: Pod spindle-shaped, small; seeds 10 to 20, crested.

Distribution: Moist woods, often forming dense carpets in Cherokee; uncommon in Iroquois; rare in Seneca.

In Kentucky: (IP-mostly), (AP-rare), (ME-rare).

This beautiful wildflower is also called blue staggers or stagger weed, which refers to the fact that the genus contains a poisonous alkaloid called cucullarine. The highest concentration of this substance is in the leaves and bulblets. Livestock grazing on this plant experience a staggering gait, convulsions, and possible death.

The Iroquois Indians made a compound infusion from the leaves that were rubbed on the limbs of runners to strengthen their bodies.

The popular name alludes to the flowers that resemble white pantaloons, with a yellow belt, hanging by the ankles on a clothes line.

Star-of-Bethlehem
Grape-hyacinth Family
Ornithogalum umbellatum L.
Hyacinthaceae

Key features: Plant poisonous; leaves basal, grasslike, in clumps; flowers white, star-shaped, with distinct green band on each tepal below.

Origin: Europe.

Life form: Perennial herb from white to brownish bulbs and bulblets.

Stems: Lacking except for flowering stalks 5 to 12 inches long.

Leaves: Blades bright green, fleshy, narrow with a pale grooved midrib; 6 to 12 inches long.

Flowers: Few, in terminal flat-topped clusters; tepals 6, white; stamens 6; pistil 1; 2 inches wide. April through June.

Fruits: Capsules 3-lobed; seeds several, tiny, black, grandular.

Distribution: Disturbed ground in turf, fields, thickets, roadsides, moist low woods, cultivated beds. Common.

In Kentucky: AP, (IP-common), ME. Listed as a Significant Threat by the Kentucky Exotic Pest Plant Society.

Also called snowdrops, summer-snowflake, and nap-at-noon, this ornamental plant was introduced from Europe and has escaped to become a troublesome weed. It reproduces by bulbs and can easily be confused with wild garlic except that they do not smell of garlic.

This species contains toxic cardiotox-ins. Cattle and other livestock can be poisoned by eating any part of the plant, either fresh or dried in hay. It is also poisonous to humans.

The plants die back soon after setting seed in early summer and disappear from the landscape, thus making weed control timely.

Another not-so-popular common name is dove's dung and suggests the similarity of a mass of white flowers to that of bird droppings.

Maple-leaved waterleaf

Waterleaf Family
Hydrophyllum canadense L.
Hydrophyllaceae

Key features: Leaves palmately 5-to-9-lobed often mottled in shades of green; flowers below the leaves white to pale purple, bell-shaped, stamens extending beyond the petals.

Origin: Native.

Life form: Perennial herb from scaly rhizomes.

Stems: Slender, erect, smooth to slightly hairy; 12 to 24 inches tall.

Leaves: Lower leaves long-stalked rising above the flowers; margins coarsely toothed, tips pointed, bases with wide sinus or v-shaped; upper leaves smaller, short-stalked.

Flowers: In dense or loose clusters, 1 to 1½ inches wide; sepals 5, linear-lanceolate, deeply cleft; petals 5; stamens 5; April through June.

Fruits: Capsule rounded; seeds 1 to 4.

Distribution: Moist damp woods, ravine slopes, often forming dense colonies along Beargrass Creek in Cherokee, where it is the dominant species.

In Kentucky: AP, IP, (ME-rare).

Overwintering in a basal rosette, in spring the new leaves and seedlings are often mottled in shades of green, or "watermarks," which aptly describes the family—*hydro* "water" and *phylum* "leaf."

White fawn-lily
Lily Family
Erythronium albidum Nutt.
Liliaceae

Key features: Plant from a corm; leaves basal, dark green, often mottled in purplish brown to pale green; flowers solitary, white, nodding from the top of a slender stem.

Origin: Native.

Life form: Perennial herb.

Stems: Lacking except for the flowering stalks to 9 inches long.

Leaves: Basal, 1 or 2, blades lanceolate to elliptic, margins entire, veins parallel, base tapering into a stalk, tips pointed; 3 to 6 inches tall.

Flowers: Tepals 6, white to pale lavender, linear-lanceolate, becoming recurved; stamens 6, anthers deep yellow; pistil 1, stigma 3-lobed; 1½ inches long. March through April.

Fruits: Capsule 3-angled; seeds flattened, in 2 rows.

Distribution: Low moist woods, slopes; prefers calcareous soils. Rare at Shawnee, uncommon at Cherokee, and locally common at Iroquois.

In Kentucky: (AP-rare), (IP-mostly inner Bluegrass), ME.

In early spring, this wildflower makes a spectacular display in low woods at Iroquois. If left undisturbed, it can produce large colonies over time. Like yellow fawn-lily, it takes several years for a plant to develop a flower, and often plants with one leaf are more abundant than plants with two leaves and a flower.

White fawn-lily was listed in the *Pharmacopoeia of the United States* from 1820 to 1863 and was used to treat gout. It was also used by the early settlers as a substitute for meadow saffron (*Colchicum autumnale*).

Bloodroot
Poppy Family
Sanguinaria canadensis L.
Papaveraceae

Key features: Plant from a rhizome; single basal leaf wrapped around the flower stalk when young; flowers white, solitary.

Origin: Native.

Life form: Perennial herb from a rhizome.

Stems: Lacking except for the flower stalk 6 to 12 inches tall.

Leaves: Blade single, orbicular in outline, 3-to-9-lobed, palmately veined; surface blue-green above, whitish green below, margins wavy-toothed; expanding to 11 inches wide.

Flowers: Sepals 2, falling early; petals 8 to 12, oblong to elliptic; stamens numerous, anthers yellow; pistil 1, green, stigma 2-grooved; 1 to 3 inches wide. March through April.

Fruits: Capsule 2-valved, slender, tapering at both ends, beaked, and opening longitudinally; seeds 10 to 15, keeled.

Distribution: Moist wooded slopes. Uncommon at Cherokee; rare at Iroquois.

In Kentucky: AP, IP.

The flower of bloodroot is short-lived, as the delicate petals easily fall off in the spring rain and wind. The leaf is wrapped around the flower bud as it emerges from the ground and doesn't fully expand until the flower has bloomed.

The Native American name, *puccoon,* may have been derived from *pak,* "blood." The bright red juice was used to paint their bodies, clothes, weapons, and baskets. It was also used as a love charm. The Iroquois Indians mixed the roots with whiskey and took it as a blood remedy for tapeworms. It was also reported to be helpful in curing rattlesnake bites. The Cherokee Indians made a decoction from the root that was used in a small dose for coughs, croup, and lung inflammations.

The genus is from *sanguinarius* "bleeding" and refers to the color of the juice.

Rue-anemone

Buttercup Family
Anemonella thalictroides (L.) Spach
Ranunculaceae
[*syn=Thalictrum thalictroides* (L.)
Eames & B. Boivin]

Key features: Upper leaves divided into 3 leaflets with 3 bluntly lobed tips and produced in a whorl below the 3 to 6 white to pink flowers; basal leaves similar but long-stalked.

Origin: Native.

Life form: Perennial herb from cluster of tubers.

Stems: Erect, slender, delicate; to 10 inches tall.

Leaves: Blades obovate to roundish, blue-green below, bases heart-shaped; ½ to 1 inch wide.

Flowers: Solitary or few in an umbel, all on threadlike stalks; sepals 5 to 10, white to pink; petals absent; stamens numer-ous; pistils several; ½ to ¾ inch wide. March through May.

Fruits: Achenes tiny, round to spindle-shaped, 8-to-10-ribbed, not beaked.

Distribution: Moist woods. Uncommon.

In Kentucky: AP, IP, ME.

One of the earliest-blooming spring wildflowers, this species is also called windflower or wind anemone because the fragile blossoms are said to blow away with the wind. The generic name honors *Anemose,* the Greek god for wind. It was believed in ancient times that if a strong wind passed over a field of anemones, it became poisoned and could cause sickness. Both the Egyptians and Persians regarded it as an emblem of sickness.

Double-flowered plants are often seen, and the first flowers that open are the largest.

False rue-anemone

Buttercup Family
Enemion biternatum Raf.
Ranunculaceae
[*syn=Isopyrum biternatum* (Raf.) Torr. &
A. Gray]

Key features: Stems delicate, reddish
green; basal leaves long-stalked, upper
alternate, compound; flowers white,
solitary or in groups of 2 or 3 in leaf
axils.

Origin: Native.

Life form: Perennial herb from fibrous
roots.

Stems: Slender, smooth, delicate, slightly
branched; 6 to 10 inches tall.

Leaves: Compound, 3-foliate, middle
leaflet on longer stalk than other two,
blades ovate to oblong, tip bluntly
3-lobed or cleft, dull green; about 1 inch
long. (New bright green basal leaves can
be seen coming up in fall.)

Flowers: Sepals 5, white, falling early;
petals absent; stamens many, anthers
yellow; pistils few, green; subtended by
bractlike leaves; ½ to ¾ inch wide. April
through May.

Fruits: Pods beaked, usually 4-valved,
elliptic to obovate, spreading; seeds few,
tiny.

Distribution: Moist woods, ravine slopes,
woodland edges, especially in limestone.
Common at Cherokee; rare in Seneca.

In Kentucky: IP.

"Rue" was a term used for several American wildflowers whose leaves resembled
those of the graceful, lacy, pungent
European herb Rue (*Ruta graveolens*),
which is still used in the landscape today.

This species is often confused with
rue-anemone (*Anemonella thalictroides*)
where they both can be seen growing
together, although false rue-anemone is
more common.

Goldenseal
Buttercup Family
Hydrastis canadensis L.
Ranunculaceae

Key features: Plant from a knobby golden yellow rhizome; leaves wrinkly, basal leaf one and a pair at the summit; flowers solitary, with many white stamens.

Origin: Native.

Life form: Perennial herb from a rhizome and fibrous roots.

Stems: Light green to reddish, hairy; 8 to 12 inches tall.

Leaves: Palmately compound, lobes 5 to 7, margins densely hairy, doubly toothed or cleft, veins hairy below, bases heart-shaped to V-shaped, tips pointed; 2 to 4 inches long.

Flowers: Stalked, from the uppermost leaf; sepals 3, greenish white, reflexed, falling early; petals absent; stamens 40 or more, filaments white, anthers yellowish green; pistils 12, greenish white, short-beaked; $\frac{1}{2}$ inch wide. April through May.

Fruits: Berries bright red, 10 to 30 in a cluster; seeds black, shiny, 1 to 2.

Distribution: Moist open woods. Rare.

In Kentucky: AP, IP, ME.

Also called orange-root because of its golden yellow rhizomes, this plant possesses many different medicinal properties that have been valued throughout history. The Iroquois Indians made an infusion with the roots and whiskey that was taken for heart trouble and healing a run-down system. Today herbalists recommend it for colitis, gastritis, duodenal ulcers, lack of appetite, and liver disease. Its bitter taste aids digestion, stimulates the appetite, and increases bile secretion.

Because of its popularity as a medicinal plant, overcollecting of the rhizomes has resulted in reduction of plant populations, and the species is monitored in some states.

Common blackberry
Rose Family
Rubus alleghaniensis Porter.
Rosaceae

Key features: Stems erect or high arching, but not rooting at the tips; leaves palmately lobed, green below; flowers white, showy, on densely glandular-hairy stalks.

Origin: Native.

Life form: Semi-woody shrub producing long canes forming loose colonies.

Stems: Both primocanes (first year's shoot, usually flowerless) and floricanes (flowering and fruiting shoots) erect or high arching, stout, angled, dull reddish brown, prickles scattered, broad-based, hooked, mostly pointing downwards; to 7 feet long.

Leaves: Alternate, leaflets 5 (3 to 7), blades ovate to oval, margins sharply toothed, bases rounded or tapering, tips pointed to short-tapering; floricanes mostly with 3 leaflets; primocanes mostly with 5 leaflets, terminal leaflet largest, stalks glandular hairy; stipules at base of leaf stalk and stem very narrow, ½ inch or less.

Flowers: Showy, 1 inch wide, in elongated clusters exserted beyond leaves, not hidden; sepals 5, green, ovate, tips elongated; petals 5, broad, wrinkly, longer than sepals; stamens and pistils many. April through June.

Fruits: Rounded or thimble-shaped, shiny dark purple to black, juicy, ripening in July through September.

Distribution: Woodland edges, thickets, fields, clearings. Common.

In Kentucky: AP, IP.

Similar species: Black raspberry (*Rubus occidentalis* L.) has **arching canes** that are reddish purple and white-tinged and **root at the tips.** The alternate leaves are **palmately compound with 3 to 5 toothed leaflets** that are green above and **white below.** The **nonshowy flowers** have 5 green sepals and 5 small, whitish green petals. Cher, Sen. Woodland edges, thickets, clearings. Common. April through June. In Kentucky: AP, IP, (ME-rare).

The roots and rhizomes of many *Rubus* species have high tannin content and have been used as an astringent. The leaves of black raspberry have been used as a wash for sores, boils, and ulcers and as a root tea for stomach pain and diarrhea.

The succulent, sweet berries contain pectin and are used in making jams and preserves, and recipes abound for making delicious pies, tarts, and other pastries.

Cleavers
Madder Family
Galium aparine L.
Rubiaceae

Key features: Stems 4-angled with tiny backward-pointing hooked bristles; leaves whorled, entire; flowers white, tiny, stalked.

Origin: Native.

Life form: Winter or summer annual herb from a taproot.

Stems: Weak, reclining to sprawling; 1 to 3 feet tall.

Leaves: In well-separated groups of 6 to 8; blades narrow, 1-nerved, margins with backward-pointing hooked bristles, midvein with hooked bristles, tips pointed; to 3 inches long.

Flowers: Single or in axillary clusters of 3 to 5 on stalks exceeding the leaves; sepals rudimentary; petals 4, tips pointed, stamens 4; styles 2; ⅛ inch wide. April through June.

Fruits: Fleshy, glove-shaped, yellowish brown to gray-brown seeds circular, densely covered with hooked bristles and warts.

Distribution: Moist woods, woodland edges, thickets, fields. Common.

In Kentucky: AP, IP, ME.

Galium means "milk." This plant was used in the early days to curdle milk for making cheese, and because the prickly stems cling together, they have been used also to make coarse sieves for straining milk.

The Iroquois Indians used this plant as a wash to help relieve the itch from poison-ivy. It was also good for making hair grow long. A pot of mutton, oatmeal, and cleavers was said to keep women from getting fat.

The weed is also called goosegrass, as it is eaten by geese and turkeys and can be collected for poultry feed.

Bent trillium

Trillium Family
Trillium flexipes Raf.
Trilliaceae

Key features: Leaves in a whorl of 3, upper surface dark green; flowers solitary, white, often nodding below the leaves.

Origin: Native.

Life form: Perennial herb from a short, thick rhizome.

Stems: Lacking except for flowering stalk to 20 inches tall.

Leaves: Blades broadly rhombic-oval, veins converging toward pointed tip, margins entire, bases wedge-shaped; 3 to 6 inches long.

Flowers: On an upright or nodding stalk; sepals 3, green; petals 3, spreading backward; stamens 6, creamy yellow; ovary white or pinkish, stigmas 3, recurved; about 2 inches wide. April through May.

Fruits: Capsule, purplish, round, 6-angled; seeds many.

Distribution: Moist rich woods. Uncommon at Cherokee but is coming back in some places where invasives were removed; rare at Iroquois.

In Kentucky: IP.

This spring wildflower often grows in small numbers and is a sight to behold when flowering in rich woods at Cherokee. Sometimes a variant is observed within the local population where the three petals are green and leaflike. The genus name is Greek for "three" and describes the number of leaves, sepals, and petals.

Navel corn-salad

Valerian Family
Valerianella umbilicata (Sull.) Alph. Wood
Valerianaceae

Key features: Stems forking above; leaves opposite, sessile or clasping, teeth few; flowers white, in flat-topped clusters; outermost green bracts smooth.

Origin: Native.

Life form: Winter annual from fibrous roots, often forming colonies.

Stems: Single, erect, light green, smooth or with hairs on the angles; 1 to 2½ feet tall.

Leaves: Basal, blades oblanceolate; stem leaves with blades oblong below, margins smooth or hairy, upper blades narrow, teeth few, midvein pale at base; to 3 inches long.

Flowers: In clusters, ½ to 2 inches wide; sepals 5, shorter than petals; petals funnel-form, white, lobes rounded, spreading; stamens 3, exserted; style 1; April through June.

Fruits: Rounded on 2 sides, flat on 1; seed 1.

Distribution: Moist fields, roadsides, waterways.

In Kentucky: (IP-mostly Inner Bluegrass), (ME-rare).

Similar species: Beaked corn-salad [*Valerianella radiata* (L.) Dufr.] is a native winter annual from a white taproot. The central, light green, hollow stem is often forked above and has opposite oblong leaves that are sessile below and clasping above, margins with short white hairs and **coarsely toothed below the middle.** The small, 4 to 12 white flowers are produced in flat-topped clusters ½ to ¾ inch wide. Below are the **green leaflike bracts with hairy margins.** Cher, Iroq, Shaw, Chick. Moist ground in fields, roadsides, waterways. April through June. In Kentucky: AP, IP, ME.

There are several species in this genus that can be difficult to correctly identify. The genus means "little Valerian."

Field pansy
Violet Family
Viola rafinesquii Greene.
Violaceae

Key features: Leaflike stipule at base of leaf stalk with middle lobe longer than others; flowers small, whitish to purple with petals twice as long as the green sepals.

Origin: Native.

Life form: Winter annual from a taproot.

Stems: Erect, often branched at the base, green turning reddish purple at maturity; 1 to 3 inches tall.

Leaves: Alternate, variable; lower blades rounded, upper ones obovate to linear-lanceolate, margins slightly toothed; stipules at base of leaf stalk, divided into segments, middle lobe narrow, long, margins entire.

Flowers: Solitary, on long slender stalks in the leaf axils; sepals broadly lanceolate, small; petals 5, unequal, purple-lined, center yellow; stamens 5, forming a ring around a club-like pistil; ½ inch long. March through May.

Fruits: Capsule 3-valved, green, smooth, shorter than sepals; seeds many, tiny, pale brown to yellowish brown.

Distribution: Mown fields, thin soil over limestone. Uncommon.

In Kentucky: AP, IP, ME.

This dainty native violet often goes unnoticed because of its small size, but a closer look reveals its beauty and resemblance to the ornamental pansy.

It is said that the English settlers in America were delighted to see so many violets. They used them as an ingredient in cosmetics and sweet waters. The herb, bound to the forehead, is said to induce sleep.

Cream violet
Violet Family
Viola striata Aiton.
Violaceae

Key features: Stems clump-forming, leafy; leaves basal, 3 or more, heart-shaped; flowers creamy white with purple lines, side petals bearded, lower short-spurred.

Origin: Native.

Life form: Perennial herb from a slender rhizome.

Stems: Slender, smooth, loosely spreading to ascending; 6 to 12 inches long.

Leaves: Blades deeply veined, margins bluntly or sharply toothed; tips pointed to rounded; ½ to 1½ inch long; leaflike stipule at base of leaf stalk narrow with lacerated teeth.

Flowers: Solitary, arising above the leaves from the upper leaf axils; sepals 5, green, narrow with small basal earlike lobes; petals 5, unequal; stamens 5; pistil 1; ½ to ¾ inch wide. April through May.

Fruits: Capsule widest at middle, smooth; seeds pale brown, tiny.

Distribution: Low moist woods, especially in alluvial soils. Rare.

In Kentucky: AP, IP, ME.

This dainty, sweet-smelling violet is a joy to behold in low moist woods and was common in 1941 as listed in Mabel Slack's master's thesis on the flora of Cherokee Park; today, a small population of plants can be found growing in only one location.

Violets as a group can be difficult to identify and are divided into two main groups. Cream violet belongs to a group that produces their leaves and flowers from the stems. Other violets belong to another group that produces their leaves and flowers directly from the underground root system.

Common arrowhead
Water-plantain Family
Sagittaria latifolia Willd.
Alismataceae

Key features: Plant aquatic from seeds or slender rhizomes with starchy tubers; leaves basal, variable, on long spongy stalks; flowers white, in whorls on erect stems to 2 feet tall.

Origin: Native.

Life form: Emergent aquatic herb.

Stems: Lacking except the flower stalk.

Leaves: Blades arrow-shaped, linear to ovate-triangular, veins parallel, curved, margins entire, tips pointed to blunt; young and submersed leaves strap-like, soon withering; 6 to 12 inches long.

Flowers: Male and female separate on same plant: male flowers above, female below; sepals 3, green, becoming reflexed; petals 3, broadly rounded; stamens yellow, numerous; pistils green, numerous; bracts 2 to 3, green, at base of each whorl; 1 to 1½ inches wide. June through September.

Fruits: Achenes 3-angled, winged with a lateral beak.

Distribution: Shallow water and mudflats along Cherokee Lake and pond in Iroquois. Rare.

In Kentucky: AP, IP, ME.

Also called duck potato and wahoo, this species was a favorite food and major source of starch among Native Americans. A botanist on the Lewis and Clark Expedition observed a woman collecting the roots near the mouth of the Columbia River. Using her toes, she separated the tuber from the root. Once it floated to the top, it was collected and put in a basket. Although no longer a popular food item, tubers are valuable to wildlife as a source of food and shelter for swans, ducks, and muskrats—hence the name swan potatoes.

Burdick's wild leek
Onion Family
Allium burdickii (Hanes) A.G. Jones
Alliaceae

Key features: Plant with strong onion odor; bulbs ½ to ¾ inch thick; leaves narrow, upright, fleshy; umbels few-flowered, creamy white, appearing in late spring.

Origin: Native.

Life form: Perennial herb from an onion-scented bulb.

Stems: Lacking except for the naked flower stalk up to 6 inches tall.

Leaves: Basal, blade lanceolate to elliptic, flat, blue-green, veins parallel, margin entire, bases nearly stalkless, tips tapering; 8 to 12 inches long, ¾ to 1¼ inches wide; develop in spring and wither before flowers appear in late spring.

Flowers: Umbels, ½ to ¾ inch wide; tepals 6, persistent, even in fruit; stamens 6, anthers pale yellow; pistil 1, greenish yellow; leaflike bracts 2, ovate, subtending each erect flower cluster. May through June.

Fruits: Capsules rounded, 3-lobed, green, in umbels; seeds 3.

Distribution: Moist wooded slopes and low moist woods. Rare, but scattered local population found growing in disturbed woods at Cherokee; locally common in low moist woods at Iroquois.

In Kentucky: (AP-rare), (IP-rare).

Also called narrowleaf wild leek, this rare leek is easily overlooked because it is similar in appearance to the common, more robust species *Allium tricoccum*, also known as wild leek or ramp.

The plant is named in honor of Dr. J. H. Burdick of the Gray Herbarium of Harvard University, who first recognized and described this species as separate from ramps.

Poison hemlock
Parsley Family
Conium maculatum L.
Apiaceae

Key features: Plant poisonous; gives off a rank odor when crushed; stems purple-mottled; leaves finely divided; flowers white, tiny, in numerous umbels.

Origin: Eurasia.

Life form: Biennial herb from a long, whitish taproot.

Stems: Erect, smooth, hollow, produced in the second year; to 7 feet tall.

Leaves: Alternate and basal; lower ones up to 20 inches long, upper progressively smaller; blades broadly triangular-ovate, 3-to-4-time pinnately divided with oblong to lanceolate toothed lobes.

Flowers: In open, mostly terminal compound umbels, 2 to 3 inches wide; sepals 5; petals 5; stamens 5; styles 2. June through August.

Fruits: Pale brown, ovoid, flattened on 1 side, rounded on the other 2, ribbed and tiny.

Distribution: Disturbed ground, especially along roadsides, waterways, fields, woodland edges, thickets.

In Kentucky: (AP-rare), (IP-common). This species has been mostly restricted to the Bluegrass Region since the 1980s but has now spread rapidly southward and westward, where it has become a troublesome weed. Listed as a Severe Threat by the Kentucky Exotic Pest Plant Council.

Extremely poisonous, this member of the Parsley Family is well known because the Greek philosopher Socrates died by drinking a decoction of this plant, which causes respiratory failure. However, he was not the only one poisoned by this method in ancient Greece; extracts were often used to execute political prisoners and criminals. Reports of poisoning today usually stem from the fact that the leaves are mistaken for those of parsley, the roots for those of parsnip, and the seeds for those of anise. Even the hollow stems fashioned into whistles have poisoned children.

Conium is derived from the Greek word *konas,* which means "to whorl about."

Honewort

Carrot Family
Cryptotaenia canadensis (L.) DC.
Apiaceae

Key features: Leaves alternate, leaflets 3, entire to deeply cleft; flowers tiny, white on unequal stalks in loose terminal and axillary umbels.

Origin: Native.

Life form: Perennial herb from a fibrous root.

Stems: Erect, smooth, branching above, light green, sometimes with a white tinge; 1 to 3 feet tall.

Leaves: Long-stalked below, clasping above; blades ovate to obovate, margins sharply toothed, tips pointed, bases rounded to wedge-shaped, tapering into a short winged stalk; 1½ to 4 inches wide.

Flowers: In umbels, these divided into 3 to 8 smaller ones; sepals green, tubular; petals 5; stamens 5, white, spreading; pistil green. May through July.

Fruits: Mericarps elongated, ribbed, tapering at both ends; green at first, maturing dark.

Distribution: Moist ground in woods, waterways, shady roadsides, limestone edges. Common.

In Kentucky: AP, IP, ME.

The generic name, *Cryptotaenia,* comes from the Greek *cryptos* "hidden" and *tainia* "a fillet," alluding to the concealed oil tubes that are found in the furrows of the primary ribs in the elongated fruit that splits down the middle.

Wild carrot
Carrot Family
Daucus carota L.
Apiaceae

Key features: Plant from a carrot-like taproot; leaves finely divided; flowers tiny, white, in flat-toped to slightly rounded umbels.

Origin: Eurasia.

Life form: Biennial or summer annual herb.

Stems: Erect, sometimes branched, ridged, hollow; 1 to 5 feet tall.

Leaves: In a basal rosette the first year, long-stalked; divided 3-times into fernlike narrow-lobed segments; alternate stem leaves above, similar, but clasping; to 6 inches long.

Flowers: Sepals 5; petals 5, white; involucral bracts dissected into very narrow segments. (There is usually one dark purple flower in the center of each umbel which blooms first). June through September.

Fruits: Divided into 2 sections, yellow-brown, oval, ridged, prickly on ribs.

Distribution: Disturbed ground, especially along roadsides, fields, thickets. Common where it escapes mowing.

In Kentucky: AP, IP, ME. Listed as a Significant Threat by the Kentucky Exotic Pest Plant Council.

Wild carrot is also known as Queen Anne's lace because the lacy leaves were often used in fashionable head-dresses and bouquets of the seventeenth, eighteenth, and nineteenth centuries.

It was first reported in the northwestern United States in the 1800s. It spread quickly from 1921 to 1960 and became established in areas where large quantities of exotic crop seeds were imported.

A plant of antiquity, the seeds have been unearthed and dated from the

Invasive plant
Cher, Sen, Iroq, Shaw, Chick

Neolithic, the Bronze Age in Switzerland, and Iron Age in Sweden. A biotype from Afghanistan with purple roots is believed to be the progenitor of the popular carrot crop of today. The popularity of the purple roots soon faded in Europe because it gave food an unappetizing brownish purple color. It was replaced by the orange carrot, which was selected in the 1600s in the Netherlands and is the popular carrot of today.

Rattlesnake-master

Carrot Family
Eryngium yuccifolium Michx.
Apiaceae

Key features: Basal and lower leaves long, stiff with needlelike teeth; flowers tiny, white, produced in numerous, rounded thistle-like heads

Origin: Native.

Life form: Perennial herb from a taproot.

Stems: Stout, bluish green, grooved, usually not branched; 2 to 4 feet tall.

Leaves: Blades narrow, blue-green, often curving downward, parallel-veined; upper leaves similar, gradually reduced; to 2½ feet long.

Flowers: Numerous, produced in dense terminal and axillary heads, 1 inch wide; sepals absent; petals 5, white; stamens several, white, anthers brown; styles threadlike, white, protruding; involucral bracts whitish gray, short to slightly protruding beyond base of flower heads. July through August.

Fruits: Capsules brown; seeds many, ¼ inch long.

Distribution: Fields, thickets at Cherokee; Summit Field at Iroquois. Uncommon.

In Kentucky: AP, IP, ME.

Looking more like a desert plant, the species name comes from the resemblance of the strap-like leaves to that of a yucca plant. The common name refers to the belief by Native Americans that an infusion made from the roots would heal rattlesnake bites. The dried seed heads were used as rattles, and the fruits and leaves were used in rattlesnake medicine song and dance. It was also used as a sedative and produced feelings of peace and tranquility.

Hemp dogbane

Dogbane Family
Apocynum cannabinum L.
Apocynaceae

Key features: Plant with milky sap; leaves opposite; flowers white or greenish, bell-shaped, in dense terminal and axillary clusters; fruit pods dangling in pairs.

Origin: Native.

Life form: Perennial herb from rhizomes.

Stems: Erect, slightly woody at base, branched above, reddish green; 2 to 4 feet tall.

Leaves: Sessile to short-stalked; blades ovate-oblong to lanceolate, dark green above, pale below, distinctly white-veined, margins smooth, bases tapered, tips short-pointed; 3 to 5 inches long.

Flowers: Sepals 5; petals 5, fused into a tube with 5 spreading triangular lobes; stamens 5, anthers orange forming a cone around the stigma; ¼ inch or less. May through August.

Fruits: Seedpods straight or curved, opening along one side, to 8 inches long; seeds brown small, tufted hairs white.

Distribution: Woodland edges, roadsides, thickets, fields. Common.

In Kentucky: AP, IP, ME.

Hemp dogbane is known for the long, brownish fibers from its stems, which are strong and durable, and used in making fish nets, string, and rope. Peter Kalm, a student of the great botanist Linnaeus, wrote that Swedish colonists along the Delaware River preferred rope made from hemp dogbane to that of the common hemp (*Cannabis sativa*) and bought yards of it from the Native Americans in exchange for a piece of bread.

Fibers from this plant have been identified in Ohio Hopewell and Adena fabrics that date back to 100–300 BC.

Sandvine
Milkweed Family
Ampelamus albidus (Nutt.) Britton
Asclepidaceae
[*syn=Cynanchum laeve* (Michx.) Pers.]

Key features: Plant twining; leaves opposite, heart-shaped; flowers small, whitish green, borne in rounded clusters in the leaf axils; fruit pods slender.

Origin: Native.

Life form: Herbaceous perennial vine from rhizomes.

Stems: Smooth, slender, without milky sap; to 16 feet long.

Leaves: Stalked, blades dark greenish blue, veins distinct, margins entire, often wavy, basal lobes rounded with wide sinus, tips long-pointed; 3 to 6 inches long.

Flowers: Calyx 5-lobed; corolla 5-lobed, center with 5 whitish green hoods, narrow, erect, divided into 2 lobes, about equaling the petals. July through September.

Fruits: Pods opening along one seam, turning brown with age, to 4 inches long; seeds many, brown, oval, winged, tipped with silky white hairs.

Distribution: Low moist woods, fields, thickets, cultivated beds, waterways. Common.

In Kentucky: IP, ME.

Often considered a troublesome weed, the strong honey-like fragrance of the flowers are popular with beekeepers, who recommend this plant as an excellent source of pollen for honey production.

This native vine is also called bluevine because of the bluish cast in the leaves. Other common names are honeyvine milkweed and honeyvine swallowort, referring to the honey-making quality of the plant and the butterfly it attracts.

White snakeroot

Aster Family
Ageratina altissima (L.) R. M. King
Asteraceae
(*syn=Eupatorium rugosum* Houtt.)

Key features: Leaves opposite, long-stalked; flower heads white, in broad terminal and axillary flat-topped to slightly dome-shaped clusters.

Origin: Native.

Life form: Perennial herb from fibrous roots and rhizomes.

Stems: Erect, simple to much-branched, smooth to slightly hairy; 1 to 5 feet tall.

Leaves: Blades ovate, lanceolate to heart-shaped, veins 3, prominent, margins sharply toothed, bases often unequal, tips pointed; to 7 inches long, smaller upward.

Flowers: Disk florets only, small, tubular, bright white, tips 5-notched, style divided extending beyond tube; involucral bracts green, sometimes white-margined, tiny. July through October.

Fruits: Achenes tiny, black or dark brown, 5-angled, narrow; 1-seeded.

Distribution: Moist to dry open woods, woodland edges, thickets, along waterways, roadsides. Common.

In Kentucky: AP, IP, ME.

Eupatorium was named by Linnaeus in honor of the Greek physician Eupator Mithridates, who made medicine from one of the species in the genus.

Early settlers believed that this plant could be used in the treatment of snake bites, but the plant is known to be highly toxic. It has long been a problem to livestock that feed on it when other forage is scarce. Cattle that graze on it may develop "trembles" or "milk sickness." The plant contains a poison called *tremetol,* which is soluble in milk fat and may be transferred to suckling animals or to humans drinking the milk from the affected animal. The symptoms include nausea, vomiting, weakness, slow respiration, and possible death. This disease is believed to have been responsible for the death of Abraham Lincoln's mother. Milk sickness was not confirmed until 1917. Another common name is milk-sickness plant.

Horseweed

Aster Family
Conyza canadensis (L.) Cronquist
Asteraceae

Key features: Stems wand-like, densely white-hairy; leaves alternate, toothed; flower heads small, whitish or greenish with a yellow center, in a loosely spreading inflorescence.

Origin: Europe.

Life form: Winter annual from a long taproot.

Stems: Erect, ridged, hairy, branched above, leafy; to 5 feet tall.

Leaves: Lower ones stalked, withering early; blades linear to oblanceolate; upper ones smaller, narrow, margins mostly entire, hairy, clasping, crowded; to 4 inches long.

Flowers: Heads numerous; inner disk florets tiny; outer ray florets white; involucral bracts slender, green, overlapping. July through October.

Fruits: Achenes straw-colored, widest at tip with tufts of white to brownish hairs.

Distribution: Disturbed ground in fields, roadsides, limestone ledges, thickets. Common.

In Kentucky: AP, IP, ME.

This species has a strong, pungent odor caused by volatile terpene oil secreted by thousands of dot-like glands. The oil may cause inflammation and irritation to the skin and to the mucous membranes of the mouth, nose, and eyes in some persons and in the nostrils and throats of grazing animals. Butter made from cattle feeding on this plant can produce a foul taste within 48 hours.

Considered a successional species, it is one of the first to invade disturbed ground. It is a prolific seed producer, and one plant can produce up to 50,000 seeds.

Other common names include butter-weed, cowtail, hogweed, and mares-tail.

Fireweed

Aster Family
Erechtites hieracifolia (L.) Raf.
Asteraceae

Key features: Stems tall, leafy; leaves alternate, sharply toothed with callous tip; flower heads numerous, whitish yellow and swollen at the base.

Origin: Native.

Life form: Annual herb from fibrous roots.

Stems: Erect, grooved, smooth to slightly hairy; to 7 feet tall.

Leaves: Lower ones oblanceolate to obovate, often stalked; middle and upper ones elliptic to lanceolate, leaf bases stalked to clasping; 2 to 8 inches long.

Flowers: Heads in elongated terminal clusters; inner disk florets enclosed by dull green to brown involucral bracts that are narrow, equal in length, and swollen at base; 1 inch long. July through September.

Fruits: Achenes tiny, ribbed, tufted hairs at apex bright white.

Distribution: Disturbed ground in open woods, fields, roadsides, cleared limestone ledges. Especially common in bare ground where invasive plants have been removed.

In Kentucky: AP, IP, ME.

This native species is weedy and one of the first herbaceous plants to become established in disturbed, cleared ground. Even in the early eighteenth century, this species was said to be one of the first to invade new plantations where the ground had been burned. The seeds, with tufts of soft white hairs, are easily dispersed by the wind.

Common quickweed

Aster Family
Galinsoga quadriradiata Ruiz & Pavon
Asteraceae
[*syn=Galingsoga ciliata* (Raf.) S.F. Blake]

Key features: Plant densely hairy; leaves opposite, short-stalked, toothed; flower heads small, daisy-like, single or few at ends of branches.

Origin: Tropical America.

Life form: Annual herb from a taproot.

Stems: Erect to spreading, freely branching, rough; 8 to 28 inches tall.

Leaves: Blades broadly ovate, 3-nerved, upper and lower surface with short white hairs, margins coarsely toothed, bases rounded, tips short pointed; 1 to 3 inches long.

Flowers: Heads less than ¼ inch wide; inner disk florets yellow, tiny, numerous; outer ray florets 5 (4–6), white, tips 3-toothed; involucral bracts green, nerved, ovate, in 2 to 3 weakly defined series. June through September.

Fruits: Achenes tiny, densely hairy, dark brown to black, widest at the tip, bristle-tipped.

Distribution: Disturbed moist to wet ground, cultivated beds, roadsides ditches, fields, pond margins, waterways. Common.

In Kentucky: AP, IP, ME.

This familiar weed occurs throughout the world and is especially troublesome in cultivated beds, vegetable gardens, and plant nurseries.

The young stems and leaves can be cooked and eaten as greens.

Other common names are fringed quickweed, Peruvian daisy, and shaggy soldier. The genus is named for Mariano Martinez de Galinsoga, a Spanish botanist of the eighteenth century.

Narrow-leaved white-top aster

Aster Family
Sericocarpus linifolius (L.) BSP
Asteraceae
(*syn=Aster solidagineus* Michx.)

Key features: Leaves narrow; flower heads white, borne in flat-topped, loose clusters; involucral bracts whitish yellow with green tips.

Origin: Native.

Life form: Perennial herb from a stout rhizome.

Stems: Single, erect, smooth; 1 to 2 feet tall.

Leaves: Alternate, lower stalked, falling early, upper ones sessile, reduced; blades firm, midvein distinct, margins entire; 1 to 3 inches long.

Flowers: Inner disk florets, white to yellowish, tiny, tubular; outer ray florets, 3 to 6, white; involucral bracts in several rows, overlapping. June through September.

Fruits: Achenes short, very silky, bristles white at tip.

Distribution: Dry upland woods. Uncommon.

In Kentucky: (AP-rare), IP, (ME-rare).

The genus name, *Seriocarpus,* is Greek—*serico*, "silky" and *carpos* "fruit"—and aptly describes the fruit.

Water-cress

Mustard Family
Nasturtium officinale R. Br.
Brassicaceae
[*syn=Rorippa nasturtium-aquaticum*
(L.) Hayek]

Key features: Plant aquatic, often forming dense mats in shallow water; stems succulent; leaves dark green, pinnately divided; flowers small, white.

Origin: Eurasia.

Life form: Aquatic perennial herb.

Stems: Smooth, creeping on mudflats or floating, bases usually submersed, rooting at the nodes, tips erect; to 18 inches tall.

Leaves: Somewhat fleshy, leaf segments 3 to 11, variable in shape from round, linear, ovate, to oblong, bluntly toothed, the terminal lobe usually the largest, margins entire.

Flowers: In dense terminal clusters; sepals 4; petals 4; stamens 6, filaments purple; pistil 1. May through August.

Fruits: Pods narrow, cylindrical, slightly curved, ascending; seeds shiny, brownish, in 2 rows.

Distribution: Spring-fed stream at Breckenridge Springs: wet roadside ditches in Cherokee. Locally common in Seneca; rare in Cherokee.

In Kentucky: IP, (ME-rare). Listed as a Significant Threat by the Kentucky Exotic Plant Pest Council.

This Eurasian plant was used for medicinal purposes and for food by the Persians, Greeks, and Romans. Cultivation began in the nineteenth century, and not only is it an important food crop throughout the world today, but it has also spread and become a cosmopolitan weed of wet areas.

If harvested before flowering, this species is rich in Vitamins A and C and iodine. The young mustard-like leaves are eaten raw or in salads and have a peppery taste. However, if collected in the wild from polluted waters, water-cress can transmit parasites, and poisoning has resulted from eating leaves that have absorbed heavy metals and toxins.

Nasturtium is from *nasus tortus,* which means "a wry or twisted nose," and alludes to the pungent flavor.

Amur honeysuckle

Honeysuckle Family
Lonicera maackii (Rupr.) Maxim.
Caprifoliaceae

Key features: Deciduous shrub with hollow branches; leaves opposite, ovate, hairy; flowers white to pinkish, turning yellow with age; berries 4, bright red.

Origin: Northeastern China and Korea.

Life form: Deciduous shrub, upright, multi-stemmed; to 20 feet tall.

Leaves: Simple, short-stalked; blades elliptical to ovate, margins entire with short hairs, surface below with short hairs mostly on the veins; tips tapering to a long slender point; 1½ to 4½ inches long.

Flowers: In axillary pairs, short-stalked; sepals green, densely hairy; petals tubular, 2-lipped, the upper lip with 4 fused lobes, lower lip single; stamens 5, anthers yellow, extending beyond the petals. May through June.

Fruits: Berries round, juicy, ¼ inch wide, turning bright red in fall; seeds few.

Distribution: Disturbed ground, especially in woods, woodland edges, thickets, roadsides, waterways.

In Kentucky: (AP-rare), (IP-common), (ME-rare). Listed as a Severe Threat by the Kentucky Exotic Pest Plant Council.

Amur honeysuckle has spread at an alarming rate and is said to be one of the worst botanical explosions of the late twentieth century along the Ohio River corridor and in east-central Kentucky. Today, this plant has invaded twenty-four states.

Native to China and Korea, this ornamental was introduced into North America in 1896 for its fragrant flowers and bright red fruit. Highly adaptable, this shrub shades out native herbs and prevents canopy tree reproduction. It colonizes by root sprouts and spreads by birds (such as starlings) and small mammals.

This species is easy to recognize in early spring because it is the first to leaf out in woods, well before other native shrubs, and is one of the last to drop its leaves in the fall.

Although this species was not listed in Mabel Slack's 1941 thesis on the flora of Cherokee Park, it is considered the worst woody invasive plant in all five parks.

The genus honors Adam Lonitzer (Latinized *Lonicerus*), a German herbalist of the sixteenth century.

Common elderberry

Honeysuckle Family
Sambucus canadensis L.
Caprifoliaceae

Key features: Woody shrub, often forming colonies by stolons; pith white; leaves pinnately compound; flowers white, in terminal flat-topped clusters.

Origin: Native.

Life form: Shrub with stems weak, arching, spreading outward away from base; to 9 feet tall.

Leaves: Leaflets 5 to 11, opposite with a single terminal leaflet, blades lanceolate to ovate, margins toothed with short hairs, veins below hairy, tips pointed, the lower occasionally divided into 3 leaflets; 3 to 6 inches long.

Flowers: In clusters 4 to 9 inches wide; sepals 5; petals 5, deeply lobed; stamens 5, protruding; style short, 3-lobed. May through July.

Fruits: Berries purplish black, juice red, with 3 to 5 small stones.

Distribution: Moist to wet ground in woods, woodland edges, waterways, thickets. Common.

In Kentucky: AP, IP, ME.

The berries produced in late summer and autumn are an important food source for songbirds and small mammals. Humans have used them for making jellies, jams, pies, and wine. The Iroquois Indians mixed the flowers in with other plants as a medicine to soak the corn seeds before planting.

The Cherokee Indians used the berries for treating rheumatism and the flowers to "sweat out" fever. Young boys used the larger stems to make popguns. However, poisoning in children has been reported from chewing or sucking the bark.

Coralberry

Honeysuckle Family
Symphoricarpos orbiculatus Moench.
Caprifoliaceae

Key features: Understory shrub, often forming large colonies; leaves opposite, round; flowers whitish pink; fruits coral to reddish pink.

Origin: Native.

Life form: Multi-stemmed with arched, spreading purplish brown branches, 2 to 4 feet tall; spreads by rhizomes.

Leaves: Blades elliptic to round, dull green above, paler below with hairs, margins entire to slightly toothed, tips blunt, rounded or sharp-pointed; 1 to 2 inches long.

Flowers: In dense axillary clusters; sepals with 5 short lobes; petals bell-shaped, 5-lobed; stamens 5; style hairy, short. July through August.

Fruits: Berry-like, round, fleshy; seeds 2.

Distribution: Open woods, woodland borders, thickets, fields, especially in limestone soils. Locally common at Cherokee; uncommon elsewhere.

In Kentucky: AP, IP, (ME-rare).

This common, often weedy shrub is found in dry, overgrazed pastures in the Bluegrass and open woods, where the bright reddish pink berries persist into winter. Also called buckbush, it is an important food plant for deer and the fruits and seeds are relished by pheasants, quail, grouse, and songbirds.

Native Americans used the charcoal from the wood for tattooing. Also, the inner bark or leaves were used as medicine for sore or inflamed eyes.

Mapleleaf viburnum
Honeysuckle Family
Viburnum acerifolium L.
Caprifoliaceae

Key features: Understory shrub; leaves opposite, 3-lobed to occasionally unlobed; flowers white, in flat-topped clusters.

Origin: Native.

Life form: Slender, twigs hairy, sparsely branched; to 6 feet tall.

Leaves: Simple, blades palmately veined, margins coarsely toothed, lower surface black-dotted, hairy, bases rounded to heart-shaped; to 5 inches long.

Flowers: Small, in clusters, 2 to 4 inches wide; sepals 5, toothed; petals 5-lobed, spreading; stamens 5. April through May.

Fruits: Elliptical to round, pulpy, crimson, ⅓ inch long, turning blackish purple in August through October, often persisting into winter.

Distribution: Dry to moist woods. Common.

In Kentucky: AP, IP.

The genus *Viburnum* contains some of the most valuable native and ornamental shrubs used in the landscape today. They are especially beautiful as an understory shrub planted in a naturalistic setting. They provide beauty year-round, and the berries are valuable to birds and wildlife.

Hedge bindweed

Morning-glory Family
Calystegia sepium (L.) R. Br.
Convolvulaceae

Key features: A vine; leaves alternate, arrow-shaped at the base; flowers white or pinkish; 2 large bracts enclose the base of the flower.

Origin: Native and from Eurasia.

Life form: Perennial vine from fibrous roots and rhizomes.

Stems: Twining or trailing, light green to red, smooth to hairy; to 10 feet long.

Leaves: Blades heart-shaped to triangular-ovate, basal lobes 2, broad, squared, margins entire; 2½ to 5 inches long.

Flowers: Solitary, long-stalked in leaf axils; sepals green, ovate to elliptic; petals 5, funnel-form; 2 to 3 inches long. June through September.

Fruits: Capsule 3-angled, 1 side rounded, 2 flat; seeds 2 to 4, grayish black.

Distribution: Disturbed ground. Common.

In Kentucky: AP, IP, ME.

Similar species: Field bindweed (*Convolvulus arvensis* L.) is a perennial herbaceous vine from Eurasia. It has alternate, **triangular to arrow-shaped leaves, 1 to 2 inches long**. The flowers, about 1 inch wide, are white or pinkish, funnel-form, long-stalked, and borne in the leaf axils. **Below each flower are minute leaflike bracts.** Cher, Sen, Iroq, Shaw, Chick. Disturbed ground. Uncommon. May through September. In Kentucky: AP, IP, ME.

In the "Historia Naturalis," Pliny the Elder, the great herbalist, likened the hedge bindweed to the lily: "For whiteness they resemble one another very much as if nature in making this flower were learning and trying her skill how to frame the lily indeed." The delicate flower resembles a cap and is often called lady's nightcap.

Field bindweed was first observed in the United States in Virginia in 1739. By the 1920s, many states west of the Mississippi River listed it as a serious weed. It is known to be allelopathic.

Wild sweet potato vine
Morning-glory Family
Ipomoea pandurata (L.) G. Mey
Convolvulaceae

Key features: A vine; leaves alternate, heart-shaped; flowers white with purple centers; sepals smooth.

Origin: Native.

Life form: Perennial vine from a taproot.

Stems: Climbing or twining, freely forking, purplish, smooth to slightly hairy; to 10 feet long or more.

Leaves: Blades heart-shaped, basal lobes 2, margins entire to slightly lobed, tips sharp pointed; to 6 inches long.

Flowers: Showy, in axillary clusters of 1 to 7 on long stalks; sepals 5, blunt-tipped; petals 5, open funnel-form; stamens 5, fused to the base of tube; 2 to 4 inches long. May through September.

Fruits: Capsules oval, enclosed by leaflike sepals; seeds blackish brown, 1 surface rounded, 2 flat, hairy on angles.

Distribution: Disturbed open ground, thickets, roadsides, fields; especially common at Shawnee.

In Kentucky: AP, IP, ME.

This species is also aptly named bigroot morning glory, man-under-the-ground, and man-of-the-earth, all references to the large edible storage roots that can weigh up to twenty-nine pounds. Although these roots are deep underground, digging them out is worth the effort because they are valuable sources of starch.

The most important crop plant in the family is sweet potato (*Ipomoea batatus*), with hundreds of varieties and cultivated throughout the tropics.

Wild yam
Yam Family
Dioscorea villosa L.
Dioscoreaceae

Key features: A vine; leaves whorled to occasionally opposite below, alternate above, heart-shaped, lacking bulbils; flowers minute, whitish green; fruits 3-winged.

Origin: Native.

Life form: Deciduous perennial vine from a thick rhizome.

Stems: Smooth, varying in color from green, reddish green, pale yellow to dark red, twining counterclockwise; to 15 feet.

Leaves: Blades heart-shaped to arrow-shaped, 2 to 4 inches long, upper surface green, smooth, lower surface hairy, margins entire, distinctly palmately veined with veins running toward tip, long-stalked.

Flowers: Male and female flowers on separate plants: male flowers minute, in clusters, 4 to 12 inches long from the leaf axils; tepals 6, stamens 6; female flowers minute, in clusters, 3 to 9 inches long from the leaf axils; tepals 6; stamens 6, infertile; about ⅛ inch wide. May through June.

Fruits: Capsules ovate, maturing golden green, about 1 inch wide.

Distribution: Moist woods. Uncommon.

In Kentucky: AP, IP, ME.

Similar species: **Chinese yam** (*Dioscorea polystachya* Turcz.) is an invasive deciduous vine with twining stems to 40 feet long. Native to China and India, this species has leaves similar to that of wild yam, but has **distinct aerial potato-like bulbils located in the leaf axils** that form in late summer and fall. Rarely flowering, it spreads by the bulbils and underground tubers and can quickly form a dense cover smothering surrounding vegetation. Cher, Iroq. Uncommon. It is being eradicated due to its invasive tendencies. Woodland edges, waterways, thickets. May through June. In Kentucky: (AP-rare), IP, (ME-rare). Listed as a Severe Threat by the Kentucky Exotic Pest Plant Council.

Both species belong in an economically important genus *Dioscorea*, or yams. With over 60 species cultivated in Southwest Asia, West Africa, and Central and South America, the edible tuber is a major food source in many countries. Potato yams have a high energy value and are used in much the same way as potatoes or sweet potatoes.

Wild yam has many uses in traditional medicine and continues to be used in modern medicine. Native Americans made a root tea to relieve labor pains, colic, morning sickness, rheumatism, and "chronic gastritis of drunkards." Some contraceptives on the market today were originally derived from wild yam.

Low-bush blueberry
Heath Family
Vaccinium pallidum Aiton
Ericaceae

Key features: Woody shrub with zigzag stems covered with tiny, white warty speckles; flowers urn-shaped, yellow stamens not protruding; berries dark blue, sweet, and edible.

Origin: Native

Life form: Low-growing, 2 to 3 feet tall, often forming colonies.

Leaves: Alternate, simple, short-stalked; blades elliptic, dull green above, paler below, margins entire, finely toothed or with short hairs, tips with needlelike point; 1 to 2 inches long.

Flowers: Several, small, in clusters; sepals green, tiny; petals whitish green, often tinged with red or pink, 5-lobed; about ¼ inch long. May through June.

Fruits: Berries small, dark; seeds many, maturing in July and August.

Distribution: Dry upland woods; especially common on the south and west facing slopes in chestnut oak/pine woods.

In Kentucky: AP, (IP-but absent from Inner Bluegrass).

Similar species: **Deerberry** (*Vaccinium stamineum* L.) is a diffusely branched native shrub growing to 5 feet tall. It has **hairy twigs that lack white warty speckles** and leaves to 4 inches long. **Flowers are bell-shaped**, about ¼ inch long, whitish green, with 5 spreading lobes shorter than the **protruding yellow stamens.** The **yellowish green plump berries** ripen in July and August and are **bitter and inedible.** Iroq. Dry upland woods with low-bush blueberry. Uncommon. In Kentucky: AP, (IP-but absent from the Inner Bluegrass).

The generic name is from the Latin word *Vaccomium,* which means "berry." The low-bush blueberry is also called dryland blueberry and upland blueberry and is a beautiful clump-forming plant valuable to wildlife. The Native Americans and early colonists boiled the berries and made a sugary sauce to eat with their meat.

Largeleaf wild indigo

Legume Family
Baptistia alba (L.) Thieret
Fabaceae
[*syn=B. lactea* (Raf.) Thieret; *B. leucantha*
Torr. & A. Gray]

Key features: Stems widely spreading, bushy; leaves alternate stalked, compound; flowers white, in long, slender, erect racemes up to 2 feet above the foliage.

Origin: Native.

Life form: Perennial herb from a stout taproot and rhizomes.

Stems: Sparsely branched, smooth, bluish green or reddish purple with a whitish cast; 2 to 4 feet tall.

Leaves: Leaflets 3, blades ovate to oblanceolate, bluish green, margins smooth, bases wedge-shaped, tips rounded; 1 to 3 inches long.

Flowers: Sepals 5; petals 5, upper petal fanlike, lateral petals 2, lower 2 straight; stamens 10; about 1 inch wide. May through July.

Fruits: Pods green, turning black at maturity, inflated, beaked; seeds many.

Distribution: Summit Field. Common.

In Kentucky: (IP-rare), ME.

This beautiful prairie wildflower is also called white wild indigo and white false indigo and has been introduced from the southeastern United States into the nursery trade. It is drought tolerant but is frost sensitive and turns black and withers on cold nights.

The genus name is from the Greek *baptizein,* meaning "to dye," and refers to the economic uses of some species that yield the color indigo.

Bundleflower

Legume Family
Desmanthus illinoensis (Michx.) MacMill.
ex B. L. Rob. & Fernald
Fabaceae

Key features: Leaves alternate, fernlike; flowers white, tiny, many in rounded clusters on long stalks in upper leaf axils; fruit pods curved or coiled.

Origin: Native.

Life form: Perennial herb from a taproot.

Stems: Multiple from base, often branched, green when young, turning brown, angled, smooth to sparsely hairy; 1 to 4 feet tall.

Leaves: Bipinnate, leaflets small, blades linear-oblong, bases unequal, tips rounded; leaflike stipules threadlike, saucer-shaped glands at the base of the leaflet; about ¼ inch long or less.

Flowers: Sepals white, 5-lobed; petals 5; stamens 5, anthers yellow, longer than petals; style white, exserted; ½ inch wide. June through August.

Fruits: In rounded clusters, smooth, coiled, green when young, dark brown when mature.

Distribution: Summit Field. Common.

In Kentucky: IP, ME. Reported to be locally abundant on the banks of the Ohio River and in several remnant native grasslands and woodlands along streams in the state.

Also called prairie mimosa, this plant with unique-looking fruit was used to cure chronic conjunctivitis by some Native Americans. One such practice was used by the Paiute Indians, who placed five seeds in the infected eye at night and washed them out in the morning. The curved seedpods were also used as rattles.

This species is often used in prairie restoration plantings and is an important food source for wildlife.

Hairy lespedeza

Legume Family
Lespedeza hirta (L.) Hornem.
Fabaceae

Key features: Stems densely hairy; leaves alternate, stalked, leaflets 3, terminal leaflet longer stalked; flowers pea-like, cream-colored with purplish red base.

Origin: Native.

Life form: Perennial herb from a rhizome.

Stems: Erect or ascending, slightly branched near tip; 1 to 3 feet tall.

Leaves: Blades obovate to orbicular, upper and lower surfaces hairy, margins smooth, veins distinct, bases rounded to wedge-shaped, tips blunt or with tiny needlelike point; to 1½ inches long.

Flowers: In short terminal racemes on hairy stalks to 2½ inches long; sepals long, narrow, hairy, tapering to a point, turning brown at maturity; petals creamy white, base purplish red; ¼ inch wide. July through September.

Fruits: Pods ovate, flat-sided, hairy, dark-veined, short-beaked.

Distribution: Dry pine-oak woods. Rare.

In Kentucky: AP, (IP-mostly outer Bluegrass), ME.

The genus is dedicated to *Vincent Manuelde C'espedes,* Spanish Governor of East Florida during the time the great botanist Michaux explored there in the eighteenth century. Later, the name was misspelled by Michaux's editor as de Lespedez.

White clover
Legume Family
Trifolium repens L.
Fabaceae

Key features: Plant creeping; leaflets 3 with pale V-shaped marking on upper surface, long-stalked; flowers many, pea-like, white or pink-tinged, in rounded heads.

Origin: Eurasia.

Life form: Perennial from a shallow-rooted taproot.

Stems: Often mat-forming, rooting at the nodes; to 15 inches long.

Leaves: Arising perpendicular along a slender stem in intervals; blades broadly elliptic to obovate, margins finely toothed, tips rounded to notched.

Flowers: In heads ½ to ¾ inch wide on long naked stalks from the leaf axils; sepals 5, narrow, triangular, ribbed, tips long; petals 5, upper one larger than two laterals; stamens 10; pistil 1. May through September.

Fruits: Pod narrow; seeds 2 to 4, small, heart-shaped, smooth, yellow.

Distribution: Disturbed ground in turf, fields, roadsides, damp open woods, cultivated beds. Common.

In Kentucky: AP, IP, ME.

Similar species: **Red clover** (*Trifolium pratense* L.) is a **clump-forming** European cultigen with nitrogen-fixing nodules on the roots. The 3 leaflets are elliptic to broadly obovate, smooth margined and with a **prominent white V** on the upper surface. The **lower leaves are stalked; upper leaves clasping and short-stalked**. **Flowers many, red to pink,** in rounded terminal heads. Cher, Sen, Iroq, Shaw, Chick. Disturbed ground in fields, roadsides, turf. Common. May through August. In Kentucky: AP, IP, ME.

White clover was first introduced into North America by the early settlers who brought it over in hay from Europe. It is one of the most important pasture legumes for livestock and is a nutritious food source for many species of upland game birds, wild turkeys, and small mammals. It is of high quality and adds nitrogen to the soil.

Red clover tea was used traditionally as a cure-all. It is also said to improve appetite and regulate the digestive tract.

The word clover comes from the Anglo-Saxon *cloefer* "club" and refers to the three-knotted club belonging to Hercules. Clover has long been an omen of good luck, especially the four-leafed clover.

Wild hydrangea

Hydrangea Family
Hydrangea arborescens L.
Hydrangeaceae

Key features: Deciduous understory shrub; leaves opposite, ovate to heart-shaped; flowers whitish green, crowded in flat or round topped clusters.

Origin: Native.

Life form: Shrub with erect, unbranched stems and brownish gray bark that peels with age, often forming colonies; 3 to 5 feet tall.

Leaves: Blade ovate to heart-shaped, simple, 2 to 6 inches long, margins toothed, tips sharp pointed, on stalks 2 to 5 inches long.

Flowers: Clusters to 6 inches wide; small, fertile flowers in center: sepals minute; petals 5; stamens 8 to 10; pistil 1. June through July.

Fruits: Capsule ribbed; seeds many, tiny, flattened.

Distribution: Moist slopes in open woods, ledges. Uncommon.

In Kentucky: AP, IP, ME.

Also called sevenbark, this attractive native shrub has a tendency for the bark to peel off in several layers, exposing various shades of gray and brown.

The Cherokee Indians made a tea from the inner green bark to stop vomiting in children; chewed bark was used for stomach troubles and high blood pressure.

The genus name is from the Greek *hydor* "water" and *aggeion* "vessel," referring to the shape of the capsule.

Rose-mallow

Mallow Family
Hibiscus moscheutos L.
Malvaceae

Key features: Flowers showy, funnel-shaped, creamy white with a reddish purple center and many yellow stamens united into a column; leaves alternate, usually unlobed.

Origin: Native.

Life form: Perennial herb from a taproot.

Stems: Stout, few or many, light green; 3 to 6 feet tall.

Leaves: Blades ovate to lanceolate, palmately veined, lower surface velvety with star-shaped hairs, margins bluntly toothed, bases rounded to broadly tapered, tips pointed; leaflike stipules narrow, at the base of leaf stalk.

Flowers: Few, on terminal stalks; sepals 5, greenish yellow, ovate, united at the base, enlarging as matures; petals 5; stamens numerous; pistil 1, stigma 5-lobed; involucral bracts linear, about 12, curved upward; 4 to 8 inches wide. July through September.

Fruits: Capsules rounded, short-beaked; seeds dark brown, plump, arranged in a ring.

Distribution: Pond margins, swamps. Uncommon.

In Kentucky: AP, IP, ME.

Similar species: Smooth rose-mallow (*Hibiscus laevis* All.) is a native perennial herb that grows to 5 feet tall. The **alternate leaves are variable, but most are arrow-shaped** with the middle lobe larger than the two side lobes. The showy funnel-shaped flowers, 3 to 4 inches wide, have **pale pink to white petals and a purplish red center.** Pond margins, swamps. Cher, Iroq, Uncommon. July through September. In Kentucky: AP, ME. (*syn=Hibiscus militaris* Cav.)

The genus is the largest in the family and contains some 300 species, most of them in tropical countries. Economically, the family is important; many members are used in the horticultural trade and for crops, such as cotton.

The Native Americans made a tea from the roots of rose-mallow that was used to cure syphilis, typhoid, worms, and stomach aches; they also made a tea of the leaves for colds, fevers, headaches, and croup. It was also an excellent nerve tonic.

The showy flowers of *Hibiscus* last only a day.

American lotus

Lotus Family
Nelumbo lutea (Willd.) Pers.
Nelumbonaceae

Key features: Aquatic perennial plant; leaves large, round, leaf stalk attached to underside in center; flowers showy, white to pale yellow.

Origin: Native.

Life form: Emergent aquatic plant from spongy rhizomes and tubers below the mud.

Stems: Stiff, stout; to 8 feet long.

Leaves: Floating leaves flat, satiny blue green above, pale green below, veins distinct, margins entire; emergent leaves similar, margins curved upward making a saucer-shaped depression; to 2 feet wide.

Flowers: Solitary on long stalks rising above the leaves; sepals and petals separate, numerous, overlapping, increasing in size inwardly, intergrading into numerous petal-like stamens, filaments slender, pale yellow, anther tip curved; 4 to 12 inches wide. May through June.

Fruits: Nutlike, with 1 seed embedded in a flat-topped receptacle maturing hard, turning blackish blue.

Distribution: Pond.

In Kentucky: (AP-rare), (IP-rare), (ME-rare). Collected from 8 counties, including Jefferson.

This spectacular aquatic plant found in shallow still waters is also known as alligator corn and pond-nuts because the thick acorn-like seeds are edible and eaten by wildlife and humans. Native Americans would use the cracked seeds, freed from the shells, in meat, soup, or roasted.

Economically, this family has many species that are cultivated for water gardens because of their beauty, and the dried fruit are used in the floral industry.

Chinese privet
Olive Family
Ligustrum sinense Lour.
Oleaceae

Key features: Shrub multi-stemmed, often forming dense thickets; leaves opposite, stalks short, hairy; flowers white, stamens 2, exserted beyond corolla tube.

Origin: China.

Life form: Shrub, densely branched and spreading, 10 to 15 feet tall, with densely hairy stems; from an extensive root system and suckers.

Leaves: Blades elliptic to oblong, upper surface smooth, dark glossy green, midveins below hairy, margins smooth, tips rounded to very short-pointed; ½ to 2 inches long.

Flowers: In terminal and axillary clusters; sepals 4, fused, cuplike; petals 4, lobes spreading, basally fused, the tube equal to, or shorter than the expanded lobes; stamens 2; pistil 1. May through June.

Fruits: Berry-like, round, dark blue, turning black when mature, hanging on through winter; seeds 1 to 4.

Distribution: Disturbed ground in woods, woodland edges, thickets, roadsides.

In Kentucky: AP, IP, ME. Listed as a Severe Threat by the Kentucky Exotic Pest Plant Council.

Similar species: Japanese privet or **border privet** (*Ligustum obtusifolium* Siebold & Zucc.) is a multi-stemmed shrub to 12 feet tall. The twigs, sepals, and short flower stalks are hairy and the ill-scented white flowers are borne in nodding clusters to 1½ inches long. The corolla tube is 2 to 3 times longer than the lobes, and the **anther bases are included within the tube.** Cher, Iroq. Roadsides, woodland edges, thickets. Uncommon. June. In Kentucky: (AP-rare), (IP-mostly east-central), (ME-rare).

There are several species of privet that have become invasive in natural areas today and are difficult to tell apart. In 1941, Mabel Slack wrote in her master's thesis on the "Flora of Cherokee Park" that common privet (*Ligustrum vulgare* L.)

was "common in thickets" in the park, but Chinese privet was not mentioned. Although both species are undercollected, Chinese privet is found throughout the southern United States and west to eastern Texas. It appears to be gaining ground and expanding its range northward throughout Kentucky, where it is troublesome in woodlands. Common privet is more widespread throughout the United States but only found in a few counties in Kentucky.

Both species were brought over as a garden ornamental. Common privet is believed to have been introduced into the United States in the 1700s, and Chinese privet in 1952.

Ragged fringed orchid

Orchid Family
Platanthera lacera (Michx.) G. Don
Orchidaceae

Key features: Leaves alternate, leafy above; flowers creamy white, the lowest petal divided into 3 lobes that are deeply irregularly cleft or fringed.

Origin: Native.

Life form: Perennial herb from slightly thickened roots.

Stems: Lacking except for the flowering stalk.

Leaves: Alternate, lower few, to 6 inches long, blades elliptic to oblong, margins entire; upper leaves crowded, narrow, smaller.

Flowers: In a loose or dense terminal cluster atop a 2-foot stem; sepals 3, green, with 2 narrow lateral ones curling under lip of lower fringed petal; petals 3, creamy-white; 1 inch long. June through August.

Fruits: Capsule cylindrical, ridged; seeds minute.

Distribution: In moist ground, Summit Field. Rare, only a few plants seen after a recent burn. A new record for Jefferson County.

In Kentucky: (AP-rare), IP, ME.

The Orchid Family is one of the largest in the world, with over 30,000 species. Orchid is from the Greek word *orchis* "testicle" and refers to the shape of the underground tuber-like structure in some terrestrial species.

Key features: Leaves mostly basal, grasslike and present at flowering time; flowers tubular, white with yellow throat, spirally arranged on slender stem.

Origin: Native.

Life form: Perennial herb from thick spreading, fleshy roots.

Stems: Erect, hairs nonglandular; 10 to 24 inches tall.

Leaves: Basal, 4 to 5, blades narrow, margins slightly curved inward, tapering at both ends, sheathing below; alternate stem leaves few, reduced upward; to 10 inches long.

Flowers: Many, small, in a densely flowered terminal spike; sepals 3, white; petals 3, white, lateral 2 narrow, lip ovate in general outline, broadest just below middle. July through August.

Fruits: Capsules ellipsoid, tiny.

Distribution: Summit Field. Rare.

In Kentucky: AP, IP, ME.

Although rare at Iroquois, coming across spring ladies-tresses amid the prairie grasses and wildflowers is an exciting find. Of the nine species of *Spiranthes* found throughout Kentucky, they are all similar to the above species and characterized by flowers more or less spirally arranged on a slender stalk. The generic name, *Spiranthes,* is Greek from *speira* "spiral" and *antho* "flower."

The common name, ladies-tresses, is a corruption of "ladies traces"; the twisted stalks were said to resemble the silken laces or garments that women would wear around their waists long ago.

Spring ladies-tresses
Orchid Family
Spiranthes vernalis Engelm. & A. Gray
Orchidaceae

Pokeweed

Pokeweed Family
Phytolacca americana L.
Phytolaccaceae

Key features: Plant tall, succulent; leaves alternate, entire; flowers whitish green, in long narrow clusters; berries dark purple.

Origin: Native.

Life form: Perennial herb from a fleshy taproot.

Stems: Stout, branching above, smooth, green to reddish purple; 3 to 8 feet tall.

Leaves: Blades oblong-lanceolate to oval, veins prominent, tips pointed; to 16 inches long.

Flowers: Many, small, produced in clusters 3 to 6 inches long opposite the leaves; sepals 5, ovate; petals absent; stamens usually 10; pistil 1. May through August.

Fruits: Berries round, compressed, juice crimson; seeds 10, round, glossy, black.

Distribution: Disturbed ground in moist woods, woodland borders, thickets, fields, roadsides, waterways. Common.

In Kentucky: AP, IP, ME.

Pokeweed is both a medicinal and poisonous plant. The roots and seeds are poisonous and contain saponic glycosides and certain types of proteins that cause the white blood cells to divide. The leaves are also poisonous, but if boiled repeatedly in water the toxin is steeped out and can be poured off. The tincture of pokeweed extract was once sold in drugstores as one of the best remedies for reducing caking and swelling of cow udders, a disease known as "garget."

One of the first natural inks used by the settlers in the New World was obtained from the fruits of this plant. Native Americans used berries to make a red dye to decorate their belongings, leading to another common name, red-ink plant.

Dock-leaved smartweed

Smartweed Family
Polygonum lapathifolium L.
Polygonaceae

Key features: Stems reddish green with swollen joints; leaves alternate, narrow, with papery sheath at leaf base; flowers whitish green to whitish pink with anchor-shaped vein tips.

Origin: Europe.

Life form: Annual herb from fibrous roots.

Stems: Erect to ascending, often branched; to 3 feet tall.

Leaves: Blades lanceolate to elliptic, margins entire, tips tapering to a long point; up to 8 inches long; papery sheath margins unequal, hairless, nerved with the membranous portions between often turning brown, tearing loose with age.

Flowers: In dense terminal and axillary elongated, drooping clusters; sepals 5, smooth; petals absent; stamens 6; style 2-parted. June through October.

Fruits: Achenes dark brown, oval to rounded, 1 or both sides flat or concave.

Distribution: Moist to wet ground, especially along Beargrass Creek where it is spreading. Common.

In Kentucky: (AP-rare), IP, ME.

Evidence from archeological sites from the Iron Age in Denmark and the Bronze Age in England and Holland has shown that the seeds were used as a food source.

Native to Europe, this annual plant is now widespread throughout the temperate regions of the world and is a major global weed. It grows in moist soil along waterways and studies have shown that the fresh seeds can float in water for up to six months.

Dotted smartweed
Smartweed Family
Polygonum punctatum Elliott
Polygonaceae

Key features: Leaves alternate, narrow with papery sheath at leaf base; flowers tiny, whitish or greenish white, glandular-dotted.

Origin: Native.

Life form: Perennial herb from vigorous rhizomes.

Stems: Erect, ascending, or reclining, simple or branched, often rooting at the lower nodes; up to 3 feet tall.

Leaves: Blades lanceolate to elliptic, tapering at both ends, margins entire; 1 to 5 inches long; papery sheath cylindrical, the summit fringed with a few long, thin bristles.

Flowers: In terminal clusters, 2 to 6 inches long, the lower flowers remote and not overlapping; sepals 5, with dark- or pale-colored glandular dots; petals absent; stamens 6 to 8, style 2-to-3-parted; both stamens and styles not extending beyond the tightly folded sepals. June through September.

Fruits: Achenes tiny, oval, black, smooth, shiny.

Distribution: Wet or moist ground in open woods, fields, thickets, meadows, waterways. Common.

In Kentucky: AP, IP, ME.

Dotted smartweed was used by the Native Americans to stop stomach, joint, and leg pain. The Iroquois Indians made a compound decoction to help with psychological problems.

In South America, it is one of the plants used by fishermen to stupefy fish for easy capture.

Origin: Native.

Life form: Perennial herb from a knotty rhizome.

Stems: Upright, single or few, usually not branched; to 3 feet tall.

Leaves: Blades ovate to elliptic, margins entire, bases rounded to short-tapered, tips pointed; 3 to 6 inches long.

Flowers: One to 3 in a bundle, short-stalked, on long, slender terminal and axillary racemes; tepals 4; stamens 4; style 2-parted, hooked. June through October.

Fruits: Achene lens-shaped, ovate, brown, shiny with 2 persistent styles.

Distribution: Moist to dry woods, woodland edges, waterways. Common.

In Kentucky: AP, IP, ME.

Jumpseed
Smartweed Family
Polygonum virginianum L.
Polygonaceae
[*syn=Tovara virginiana* (L.) Raf.]

Key features: Leaves alternate, short-stalked with papery sheath at leaf base; flowers small, whitish green to pink, widely spaced on lower part of stem becoming closer or overlapping above.

This summer wildflower is sometimes placed in the genus *Tovara,* which was named for Simon de Tovar, a Spanish physician of the sixteenth century. The common name, jumpseed, refers to the style tips in fruit that "jump" when touched.

A hot infusion made from the leaves and mixed with the bark of honeylocust was used by the Cherokee Indians as an aid for whooping cough.

White baneberry

Buttercup Family
Actaea pachypoda Elliott.
Ranunculaceae
[*syn=Actaea alba* (L.) Mill]

Key features: Plant poisonous; leaves alternate, pinnately compound, divided 2 to 3 times; flowers with many white stamens; fruits resembling a porcelain doll's eye.

Origin: Native.

Life form: Perennial herb from fibrous roots.

Stems: Erect single to slightly branched, smooth; 1½ to 2 feet tall.

Leaves: Long-stalked below, shorter above; leaflets lanceolate to ovate, margins coarsely to shallowly toothed and cleft; to 4 inches long.

Flowers: Many, in dense terminal cylindrical clusters, 1 to 3 inches long on a leafless stalk arising above the leaves; sepals 3 to 5, falling early; petals 4 to 10, minute; stamens numerous; pistil 1,

stigma 2-lobed; ¼ inch wide. May through June.

Fruits: Berries white, showy, with a black spot where stigma was attached; stalks short, thick, maturing red.

Distribution: Moist woods. Rare in Iroquois; uncommon in Cherokee.

In Kentucky: AP, IP.

This graceful wildflower has several common names. Doll's eye describes the white, rounded berries with a distinct dark circle that resembles the porcelain eye of a china doll. Another name, snakeroot, refers to the belief held by Native Americans that an extract from this species was a valuable remedy against snakebites, especially rattlesnakes. It was grouped with other plants known as "rattlesnake herbs."

Children have been poisoned by eating the attractive white berries. The ancient English word *bane* means "slayer" or "murderer" and describes the toxic qualities of this plant.

Tall anemone
Buttercup Family
Anemone virginiana L.
Ranunculaceae

Key features: Leaves basal below, upper divided into 3 to 5 deeply cut segments; flowers solitary, white to greenish white, on long stalks in upper leaf axils; fruits in "thimble-like" heads.

Origin: Native.

Life form: Perennial herb from rhizomes.

Stems: Erect, hairy, little-branched; 1 to 2 feet tall.

Leaves: Basal, long-stalked, strongly veined; upper usually in whorls of 3, divided into 3 to 5 deeply rhombic-obovate segments, terminal segment largest, margins variously cleft, lobed and toothed; to 5 inches long.

Flowers: Sepals 5, tips pointed, margins enrolled slightly; petals absent; stamens many, anthers yellowish green, filaments greenish white, of varying length; about 1 inch wide. June through August.

Fruits: Heads cylindrical, to 2 inches long; achenes densely hairy.

Distribution: Woodland edges at Cherokee; oak-savana at Summit Field. Uncommon.

In Kentucky: AP, IP.

Also called thimbleweed or nimbleweed, with reference to the fruiting head shaped like a thimble, this species has been used in Native American medicine for a variety of uses. The Iroquois Indians used an infusion of the stems and roots as a love medicine; if it were placed under a pillow, a dream would tell of infidelity. Also a mixture of crushed plants was apparently effective in removing witch-craft. Other Native Americans would direct the smoke from the seedpod up the nostrils of sick or unconscious patients to revive them.

Black cohosh

Buttercup Family
Cimicifuga racemosa (L.) Nutt.
Ranunculaceae

Key features: Plant tall; leaves large, compound with 3 equal divisions; flowers with many showy white stamens produced in narrow, erect, cylindrical clusters to 1 foot long.

Origin: Native.

Life form: Perennial herb from fibrous roots and rhizomes.

Stems: Erect, light green with purple nodes, smooth; 3 to 6 feet tall.

Leaves: Leaflets 20 or more, blades oblong to ovate, to 4 inches long, margins coarsely to sharply toothed or lobed, green above, silvery green below, bases wedge-shaped to somewhat heart-shaped.

Flowers: Sepals petal-like, insignificant, falling early; petals absent; stamens many, filaments white, threadlike; pistil 1, stigma white; ½ to ¾ inch wide. June through August.

Fruits: Pods several, split open on 1 side to release several seeds.

Distribution: Moist woods. Rare.

In Kentucky: AP, IP, (ME-rare).

This beautiful summer wildflower has several common names, including black bugbane, black snakeroot, and fairy candle. It has a fetid odor and is pollinated by carrion flies, gnats, and beetles. The disagreeable odor keeps many insects away—hence the generic name that comes from the Greek *cimex* "bug" and *fugure* "to drive away."

The Cherokee Indians made an infusion that was taken for tuberculosis and as a cough remedy. The roots were used by the Iroquois Indians to help rheumatism and as an aid to promote the flow of milk in nursing mothers. Today, it is marketed to women to help with gynecological problems.

Yam-leaved clematis

Buttercup Family
Clematis terniflora DC.
Ranunculaceae
(*syn=Clematis dioscoreifolia* H. Lev. &
Vaniot; *Clematis paniculata* Thunb.)

Key features: A woody vine; leaves opposite, leaflets usually 5, margins entire; flowers white, in loose clusters; fruits with silvery white, plumelike tail.

Origin: Asia.

Life form: Perennial vine from rhizomes.

Stems: Climbing or trailing to 30 feet long.

Leaves: Leaflets with the terminal leaflet on longer stalk than laterals; blades trian-gular ovate to broadly ovate, shiny green above, paler below, veins 3 to 5, bases rounded to heart-shaped, tips pointed; 1 to 2½ inches long.

Flowers: Numerous, showy; sepals 4, white to pale yellow, spreading, margins and lower surface hairy; petals absent; stamens numerous; pistils numerous; 1 inch wide. July through September.

Fruits: Achenes flattened; seed 1, brown, with style lengthening, silvery white, plumelike.

Distribution: Woodland edges, fields, meadows. Uncommon, but seems to be spreading.

In Kentucky: (AP-rare), IP. Listed as a Severe Threat by the Kentucky Exotic Pest Plant Council.

This showy vine, also known as leather-leaf clematis and Japanese clematis, has escaped from cultivation and has become a troublesome weed in the southern United States, where it invades woodland edges, meadows, and roadsides. It is often confused with Virgin's-bower (*Clematis virginiana* L.), which is the native species. However, the native species has coarsely toothed leaf margins and flowers that are either all male or all female.

The genus exhibits unique fruits that are easy to recognize in the field; the styles lengthen after pollination into a feathery plume that allows for the many tiny fruits to be dispersed by the wind.

White avens

Rose Family
Geum canadense Jacq.
Rosaceae

Key features: Leaves mostly with 3 leaflets, toothed; flowers small, white, solitary or few at the top of the stem; fruits bristly.

Origin: Native.

Life form: Perennial herb from rhizomes.

Stems: Slender, smooth to slightly hairy below; 1 to 2½ feet tall.

Leaves: Basal long-stalked; 3-foliate, undivided or with 3 to 5 obovate segments, margins lobed and tooth; middle stem leaves alternate, short-stalked, 3-foliate, segments oblong-lanceolate, margins sharply toothed; uppermost narrow, smaller.

Flowers: Sepals 5, becoming reflexed; petals 5, obovate, about as long as the sepals or surpassing them; stamens many; pistils numerous; bracts below sepals tiny, green; ½ inch wide. May through July.

Fruits: Heads rounded, densely hairy, bristly-hooked.

Distribution: Dry to moist woods, woodland edges, thickets, fields. Common.

In Kentucky: AP, IP, ME.

Early settlers would dig up the roots before the stem sprouted and put them in ale to enhance the flavor and prevent it from going sour. Boiled and infused with wine, it was said to make a tasty beverage that was good for treating disorders of the stomach.

The hooked seedpods can travel far and are dispersed by attaching themselves to animal fur or human clothing.

Midwestern Indian-physic

Rose Family
Porteranthus stipulatus (Muhl. ex Willd.)
Britton
Rosaceae
[*syn=Gillenia stipulata* (Muhl.) Baill.]

Key features: Stems light green; leaves divided into 3 narrow leaflets but appearing as 5 due to lower 2 stipules; flower petals white, slightly twisted.

Origin: Native.

Life form: Perennial herb from a rhizome.

Stems: Erect, many-branched, hairy; 2 to 3 feet tall.

Leaves: Alternate, lower ones pinnately divided into small, narrow, sharply cut lobes; middle and upper leaflets 3, blades lanceolate, sharply toothed, tapering at both ends; 2 to 3½ inches long.

Flowers: Few, in loose terminal and axillary clusters, long-stalked; sepals 5, short-pointed; petals 5, slender, spreading; stamens 20, filaments white, anthers light brown; pistils 5; up to ½ inch long. May through July.

Fruits: Pods 5; seeds 2 to 4, flattened.

Distribution: Dry to moist woods, especially in chestnut oak/hickory woodlands. Common.

In Kentucky: IP, ME.

The heavily dissected lower leaves on this graceful wildflower look like miniature evergreen trees when they first appear in late spring.

Also called Indian physic, Native Americans used this plant in minute doses for several ailments including swelling, rheumatism, and bee stings, but in large doses this species is potentially toxic and produces dizziness, fever, and vomiting.

The genus was named after an early nineteenth-century Pennsylvania botanist, Thomas Conrad Porter.

Multiflora rose

Rose Family
Rosa multiflora Thunb.
Rosaceae

Key features: Shrub with prickly stems; leaves alternate, pinnately compound; flowers white to pale pink; rose hips red.

Origin: Japan and Korea.

Life form: Shrub sprawling, climbing or erect with slightly arching stems; to 8 feet tall.

Leaves: Leaflets 7 to 11, blades oblong to obovate, margins sharply toothed, bases rounded, tips pointed to blunt; ¾ to 1½ inches long; stipule at leaf base fringed with threadlike teeth.

Flowers: In dense pyramidal clusters; sepals 5, green ovate-lanceolate, densely hairy; petals 5, ovate; stamens yellow, numerous; styles long exserted; ½ to 1 inch wide. May through June.

Fruits: Globe-shaped; seeds straw-colored, many.

Distribution: Disturbed ground in open woods, woodland edges, thickets, fields, roadsides.

In Kentucky: AP, IP, ME. Listed as a Severe Threat by the Kentucky Exotic Pest Plant Council.

This ornamental was introduced into the eastern United States from Japan in 1866 as rootstock for roses. It was also used in the 1930s by the U.S. Soil Conservation Service, who promoted it as a "living fence" to provide privacy for farm property as well as for erosion control and wildlife cover.

This shrub can form impenetrable thickets that provide dense shelter and food for rabbits, birds, and small mammals. Research has shown that a single plant can produce a million seeds that can survive in the soil for up to twenty years. Birds are the primary means of dispersing the seeds. It is a noxious weed in many states, including Kentucky.

Virginia buttonweed
Madder Family
Diodia virginiana L.
Rubiaceae

Key features: Stems often trailing and rooting at the nodes; leaves opposite, with 1 to 3 long bristles at base; flowers 1 to 2, white, sepals 2, persistent in fruit.

Origin: Native.

Life form: Perennial herb from a branched taproot.

Stems: Erect, ascending or trailing, branched, 4-angled, red-tinged; to 18 inches tall.

Leaves: Blades elliptic to oblong-lanceolate, margins entire to slightly toothed, tips pointed; 1 to 3 inches long.

Flowers: In the leaf axils; sepals 2, margins hairy, enlarging after flowering; petals tubular, lobes 4, spreading, densely hairy inside; stamens 4; style 1, stigmas 2; ¹/₂ inch wide. June through September.

Fruits: Capsule oval, ribbed, hairy, topped by 2 persistent sepals; seeds 2, oval-oblong, ribbed.

Distribution: Wet to moist ground, fields, turf. Common at Summit Field in Iroquois; rare in Cherokee.

In Kentucky: AP, IP, ME.

Similar species: Rough buttonweed or **poorjoe** (*Diodia teres* Walter) is a native summer annual with stiff, narrow, clasping opposite leaves. The **stipules are long and bristlelike.** There are **1 to 3 white to pale purple, tubular flowers** in the leaf axils and the **fruit capsules are topped by 4 persistent sepals.** Dry to wet ground, especially at Summit Field. Common. June through September. In Kentucky: AP, IP, ME.

There are several common names for these species. Poorjoe refers to the preferred habitat of poor soil. Button-weed describes the shape of the button-like fruits. Both species tolerate close mowing and are troublesome lawn weeds. The genus name, *Diodia,* means "thoroughfare" and refers to these weedy plants growing along roadsides.

Shiny bedstraw

Madder Family
Galium concinnum Torr. & A. Gray
Rubiaceae

Key features: Stems weak; leaves in whorls of 6; flowers tiny, white or yellowish white, on stalks 2 to 3 times forked; fruits smooth.

Origin: Native.

Life form: Perennial herb from fibrous roots.

Stems: Slender, erect or reclining, branched, smooth or with backward-pointing hairs on the angles; 8 to 18 inches tall.

Leaves: Mostly in whorls of 6 (4); blades narrow, 1-nerved, margins with minute upward-pointing hairs, bristle-tipped; ¼ to ¾ inch long.

Flowers: In loose terminal and upper axillary clusters; sepals absent; petals 4; stamens 6; styles 2. May through August.

Fruits: Rounded, tiny, 2, separating when ripe, tawny.

Distribution: Moist to dry woods, woodland edges. Common.

In Kentucky: (AP-rare), IP, (ME-rare).

The species name is from *concinnus,* "skillfully put together, beautiful, elegant or striking," and aptly describes this woodland wildflower. It belongs to a large family that is mostly subtropical and tropical and contains economically important plants like coffee and quinine as well as many ornamental plants such as gardenia, sweet woodruff, bluets, and pentas.

Lizard's-tail

Lizard's-tail Family
Saururus cernuus L.
Saururaceae

Key features: Plant aquatic; leaves heart-shaped; flowers white, crowded on the upper portion of wand-like spikes that nod at the tip.

Origin: Native.

Life form: Perennial herb from creeping rhizomes, often forming large colonies.

Stems: Erect, jointed, simple to forking above; 1 to 3 feet tall.

Leaves: Alternate with long basally sheathing stalks; blades dark green above, paler below, palmately veined, margins entire, tips pointed; 2½ to 4 inches long.

Flowers: Many, tiny; sepals and petals absent; stamens 6 to 8, filaments white, long, slender, surpassing the pistils; involucral bracts small. May through August.

Fruits: Fleshy, somewhat rounded, strongly wrinkled; seed 1.

Distribution: Moist to wet ground, mudflats, shallow water. Uncommon, but seems to spreading along Beargrass Creek.

In Kentucky: AP, IP, ME.

This beautiful native wetland species can form extensive colonies by way of aromatic underground rhizomes. The common name refers to the resemblance of the nodding spike to that of a "lizard's tail." It is also called water-dragon, swamp-lily, and breast-weed. The latter name comes from a folk remedy where a poultice was made from this plant to relieve painful breasts.

Origin: Native.

Life form: Perennial herb from short rhizomes.

Stems: Erect, light green to purplish, smooth or with lines of hairs; 2 to 3 feet tall.

Leaves: Basal rosette; blades oblanceolate to elliptic, tips pointed, stalks winged; stem leaves opposite, 5 to 7 pairs, upper smaller, fewer, clasping, blades lanceolate, margins entire to finely toothed, tips pointed; to 6 inches long.

Flowers: Terminal, in loose clusters; sepals 5, 2-lipped; petals tubular, 2-lipped, upper lip 2-lobed, lower lip 3-lobed; stamens 5; ¾ to 1 inch long. May through July.

Fruits: Capsule oval; seeds many, small, gray.

Distribution: Locally common at Summit Field; rare at Cherokee.

In Kentucky: IP, ME.

Similar species: Ashy beardstongue [*Penstemon canescens* (Britton) Britton] is a perennial herb with **hairs on the stem, midvein, and lower leaf surface.** The **narrow throat** of the flower is **strongly purple-ridged,** and the lower lip protrudes out further than the upper lip. Cher. On limestone ledges, rocky open woods along Beargrass Creek. Rare. May through June. In Kentucky: AP, IP. (*syn=Penstemon brevisepalus* Pennell)

Foxglove beardstongue

Figwort Family
Penstemon digitalis Nutt. ex Sims
Scrophulariaceae

Key features: Stems shiny, smooth; leaves basal and opposite above, glossy green; flowers 2-lipped, whitish lavender, throat opening wide, slightly purple-lined inside.

The genus name is from *pente* "five" and *stemon* "stamen" and refers to the four fertile stamens and one fuzzy sterile one. Long-tongued bees visiting the flower are guided into the floral tube by the darker purple lines. Once inside, the insect brushes against the stamens and the sticky pollen is deposited on the back of the insect.

The Figwort Family is large with several members used as garden orna-mentals. Some of the more popular plants available in the nursery trade include snapdragons, monkey flowers, speedwells, and foxgloves.

Frogfruit
Vervain Family
Phyla lanceolata (Michx.) Greene
Verbenaceae

Key features: Stems creeping to ascending; leaves opposite, narrow; flower heads small, rounded to elongated, in axillary spikes surpassing the leaves.

Origin: Native.

Life form: Perennial herb from a slender rhizome.

Stems: Weak, greenish purple, ridged, rooting at the lower nodes; 6 to 15 inches tall.

Leaves: Blades elliptic to lanceolate, veins threadlike, whitish below, margins with 7 to 11 sharp teeth above middle of leaf, leaf bases tapering below into winged stalks, tips pointed; 1 to 2 inches long.

Flowers: Sepals small, 2-cleft; petals 5, white, pink, or blue-violet, united, 2-lipped, upper notched, lower 3-lobed; stamens 4; involucral bracts small, overlapping, margin rose pink, tips pointed; about $1/16$ inch long. May through October.

Fruits: Nutlets 2, yellowish, rounded on outer side, flat on inner.

Distribution: Moist to wet ground in low fields, roadside ditches, along waterways. Common.

In Kentucky: (AP-rare), IP, ME.

The common name refers to the preferred habitat of this small plant: moist to wet ground where frogs live.

White vervain

Vervain Family
Verbena urticifolia L.
Verbenaceae

Key features: Leaves opposite, unevenly coarsely toothed; flowers tiny, white, remotely spaced on stiffly ascending branches.

Origin: Native.

Life form: Perennial herb from a taproot.

Stems: Erect, simple or branching near the base, angled, slightly hairy; 2 to 5 feet tall.

Leaves: Blades lanceolate to oblong-ovate, bases tapering into a winged stalk, tips long pointed; 2 to 6 inches long.

Flowers: Sepals 5, triangular, unequal; petals 5, white, united; stamens 4, paired; ovary 4-lobed. June through September.

Fruits: Nutlets 4, ridged.

Distribution: Woodland edges, thickets, fields, roadsides. Uncommon, but seems to be spreading.

In Kentucky: AP, IP, ME.

The word *verbena* is derived from the Latin *verber,* which means "rod," "stick," or "stem." In some parts of the country, the stems have been used to sprinkle holy water.

There are several cultivated plants of importance in this family. The European vervain (*V. officinalis*) was once regarded as a cure-all. The popular garden lantana is also related to this species.

Late eupatorium
Aster Family
Eupatorium serotinum Michx.
Asteraceae

Key features: Leaves on slender stalks, coarsely toothed, often drooping; flower heads white, small, in flat-topped clusters.

Origin: Native.

Life form: Perennial herb from fibrous roots and rhizomes.

Stems: Erect, branched, slightly grooved with lines of white hairs; to 5 feet tall.

Leaves: Lower opposite, upper alternate; blades lanceolate to ovate, veins 3 to 5, bases tapering, tips long-pointed; to 6 inches long.

Flowers: Heads with 9 to 15 flowers; inner disk florets white, tubular, style divided, protruding; involucral bracts hairy, narrow, in 2 to 3 series. August through October.

Fruits: Achenes small, smooth, with resinous dots.

Distribution: Moist to dry ground in woods, woodland edges, thickets, roadsides, pond edges, waterways. Common.

In Kentucky: AP, IP, ME.

The species name, *serotinum,* was named by the great French botanist Francois Andre Michaux and comes from the Latin word *serotinus,* meaning "to develop late or slowly." This describes the late flowering period for this common plant, which is found in natural areas as well as agricultural sites.

Pale-flowered leaf-cup

Aster Family
Polymnia canadensis L.
Asteraceae

Key features: Plant tall; leaves large with few angular lobes; flower heads small, daisy-like.

Origin: Native.

Life form: Perennial herb with fibrous roots.

Stems: Erect to ascending, light green to reddish, densely rough-hairy above; 2 to 5 feet tall.

Leaves: Blades large, broadly oblong, pinnately few-lobed, light green above, paler below, veined, tips pointed; to 16 inches long.

Flowers: In dense terminal and axillary heads; inner disk florets pale yellow, tubular; outer ray florets white, 3-lobed at the tip or sometimes minute or lacking; involucral bracts 5, linear to ovate, densely white-hairy, $\frac{1}{2}$ to 1 inch wide. August through October.

Fruits: Achenes mottled brown, 3-ribbed and 3-angled.

Distribution: Dry to moist woods, woodland edges, thickets, roadsides. Common.

In Kentucky: AP, IP.

The Iroquois Indians used this plant to help soothe a toothache.

Stems: Erect to ascending, semi-woody, stiffly branched at base, smooth to white-hairy; to 4 feet tall.

Leaves: Alternate, densely hairy; basal and lower blades lanceolate to oval; middle ones narrow, stalkless; uppermost reduced; both basal and lower leaves often not persistent.

Flowers: Heads numerous; inner disk florets yellow, tubular, 20 to 40; outer disk florets white to pink, 15 to 35. August through October.

Fruits: Achenes light brown, slightly nerved with tufts of white bristles at tip.

Distribution: Disturbed ground in fields, roadsides, woodlands edges, thickets, pond margins, waterways. Abundant.

In Kentucky: AP, IP, ME.

Similar species: Small-headed aster [*Symphyotrichum racemosum* (Elliott) G.L. Nesom] has numerous small daisy-like flowers that are **produced along one side of the slightly recurved branches.** The leaves are reduced upward into tiny bracts. The **involucral bracts have elongated narrow green tips.** Cher, Sen, Iroq, Shaw, Chick. Moist open woods, woodland edges, fields, thickets, waterways. Abundant. August through October. In Kentucky: AP, IP, ME. (*syn=Aster vimineus* Lam., *Aster racemosus* Elliot.)

Old field aster

Aster Family
Symphyotrichum pilosum (Willd.) G.L. Nesom
Asteraceae
(*syn=Aster pilosus* Willd.)

Key features: Leaves narrow, progressively smaller upward; flower heads white to pink with yellow centers, produced on more than half of the loosely spreading branches; involucral bracts spine-tipped.

Origin: Native.

Life form: Perennial herb from a taproot.

According to myth, asters were created out of stardust—hence the old English name star-wort. These plants are also known as "goodbye meadow" because they crowd out other vegetation.

Indian pipe

Indian-pipe Family
Monotropa uniflora L.
Monotropaceae

Key feature: Plant absorbs nutrients from mycorrhizal fungi that are attached to tree roots; stems, leaves, and flowers white.

Origin: Native.

Life form: Perennial myco-heterotrophic herb.

Stems: Erect, single, waxy; 3 to 9 inches tall.

Leaves: Alternate, small, scalelike.

Flowers: Solitary, terminal, bell-shaped, nodding at first, becoming erect at maturity; sepals 2 or 4; petals 4 or 5; stamens 10; pistil 1; about ¾ inch long. July through September.

Fruits: Capsule rounded, erect; seeds many, tiny.

Distribution: Dry or moist, shady woods in thick leaf litter. Rare.

In Kentucky: AP, IP, ME.

Lacking chlorophyll, this species is also known as ghost plant or corpse plant. It obtains its food via mycorrhizae (tiny fungi) that are associated with the roots of trees. It is often found growing in shady woods where the leaf litter is thick. The nodding, waxy white flower becomes upright when the tiny wind-blown seeds start to form, but turns black and decays soon after, appearing dead.

Another common name is convulsion-root because it was said to be a good remedy for spasms, fits, and other nervous conditions.

Yellowtop

Aster Family
Packera glabella (Poir.) C. Jeffrey
Asteraceae
(*syn=Senecio glabellus* Poir.)

Key features: Plant succulent with hollow stems; leaves basal and alternate, deeply divided; flower heads bright yellow, daisy-like.

Origin: Native.

Life form: Biennial or annual herb from fibrous roots.

Stems: Juicy, smooth, ridged, hollow; to 3 feet tall.

Leaves: Variable; basal and stem leaves deeply divided, each with a large, rounded terminal lobe; stem leaves, reduced upward; 2 to 8 inches long.

Flowers: In heads at the tips of slender stems; inner disk florets yellow, surrounded by a few flaring, very narrow outer ray florets; involucral bracts green, long-tapering; about $1/16$ inch long. April through June.

Fruits: Achenes small, smooth to slightly hairy.

Distribution: Moist to wet ground. Common.

In Kentucky: (AP-rare), (IP-mostly in Ohio River counties), ME.

The generic name, *Senecio,* comes from the Latin *senex,* which means "old man," an allusion to the white hairs on the apex of the fruit. The genus also contains highly toxic pyrrolizidine alkaloids that cause liver disease and cancer in high doses. These alkaloids can enter the human food supply through crops contaminated with weeds containing them or through livestock ingesting them. The foliage, flowers, and seeds contain the alkaloids. Dried and fresh plants are equally toxic—thus hay containing yellowtop should not be used.

Running groundsel

Aster Family
Packera obovata (Muhl. ex Willd.) W.A. Weber & A. Love
Asteraceae
(*syn=Senecio obovatus* Muhl. ex Willd.)

Key features: Perennial plant spreading by rhizomes; basal leaves round, tapering to a winged stalk, often purple below; flower heads yellow, in loose clusters on slender stalks.

Origin: Native.

Life form: Perennial herb.

Stems: Erect, simple to slightly branched above, smooth, ridged, with leafy basal offshoots; to 20 inches tall.

Leaves: Variable; basal blades round, obovate or fiddle-shaped with terminal lobe the largest, margins bluntly to sharply toothed; stem leaves alternate, clasping, reduced upward, white woolly hairs at base of stalk; 1 to 3½ inches long.

Flowers: Heads small; inner disk florets yellow, 5-lobed; outer ray florets yellow; involucral bracts narrow, green, tips purple, hairs white woolly; ½ to 1 inch wide. April through June.

Fruits: Achenes brown, narrow, smooth to slightly hairy with white tufts of hairs.

Distribution: Open woods, limestone ledges. Rare.

In Kentucky: AP, IP.

The groundsels make up a large group of plants that are often cultivated for their flowers and foliage, which can create a colorful groundcover in spring.

Running groundsel is also called roundleaf ragwort, which refers to the shape of the basal leaves.

Common dandelion
Aster Family
Taraxacum officinale (L.) Weber
Asteraceae

Key features: Plant with milky juice and a thick taproot; leaves crowded in a basal rosette; flower heads bright yellow, borne at the top of hollow stalks.

Origin: Eurasia and probably North America.

Life form: Perennial herb.

Stems: Lacking except for the flowering stalk to 15 inches long.

Leaves: Blades oblong to spatulate, variously lobed from deeply toothed or incised to shallow-lobed, largest lobe at the tip; leaf base narrows into a winged stalk; 3 to 12 inches long.

Flowers: Head solitary, 1 to 2 inches wide; ray florets bright yellow, many tips notched; involucral bracts of two kinds: inner green ones narrow, pointing up; outer green ones curved downward. February through December; sometimes throughout the year.

Fruits: Achenes yellow-brown, ribbed, terminating in a long slender beak tipped with soft white hairs.

Distribution: Disturbed ground, especially in turf, roadsides, fields, cultivated beds, open woods, thickets. Common to abundant.

In Kentucky: AP, IP, ME.

This species is often said to be one of the most familiar plants known throughout the world and is a widely distributed weed of Europe, North America, and Asia. About fifty to sixty species are known in this genus. *Taraxacum* comes from the word *tarassen,* which means "disorder or confusion," or from the Latinization of the Persian word *tarashqun,* "potherb." *Officinale* means "official" or "sold in shops."

Flowers, leaves, and roots of the plant can be eaten by humans. It is rich in iron, copper, potassium, vitamins B1, B2, and C. Early herbalists regarded this weed as one of the best greens for building up the blood and curing anemia.

Coffee made from dandelions is prepared from autumn roots that are cleaned, dried, and roasted until dark brown, then ground for use. The ground roots may be mixed with coffee or used alone. Dandelion greens can also be purchased in some specialty stores.

Turnip
Mustard Family
Brassica rapa L.
Brassicaceae

Key features: Plant from a swollen turnip-like taproot; basal leaves pinnately lobed, upper leaves narrow with heart-shaped clasping bases; flowers yellow, in elongated terminal clusters.

Origin: Eurasia.

Life form: Biennial herb.

Stems: Erect, branched above, succulent, smooth, bluish green to purplish green with white tinge; to 3 feet tall.

Leaves: Basal and lower ones to 10 inches long, stalked; pinnately lobed, terminal lobe the largest, margins bluntly toothed to wavy; upper and middle leaves reduced, blades oblong-lanceolate, margins toothed or entire, bases clasping.

Flowers: Sepals 4, pale yellow or green; petals 4, yellow, obovate; stamens 4; pistil 1; flower stalks threadlike, up to 1 inch long. April through August.

Fruits: Pods slender, cylindrical, ascending, beaked; seeds tiny, dark brown, round.

Distribution: Disturbed open woods, fields, roadsides, waterways, thickets. Rare in 2005, it has spread rapidly along Beargrass Creek, where it is now common.

In Kentucky: (IP- rare, collected in 4 counties to date in Bluegrass Region).

The *Brassicaceae* is an economically important plant family with many cultivated food crops that have been harvested since ancient times. Members such as turnips, cabbage, kale, collards, cauliflower, broccoli, and mustards are just a few. They are rich in vitamin A and C, fiber, and isothiocyanates (mustard oils), which are cited by the National Cancer Institute for their cancer-preventing attributes. Other common names include field mustard or wild mustard.

Creeping yellow cress
Mustard Family
Rorippa sylvestris (L.) Bessr.
Brassicaceae

Key features: Plant from rhizomes often forming large colonies; leaves basal and alternate, divided; flowers bright yellow, small, in terminal and axillary clusters.

Origin: Eurasia.

Life form: Perennial herb.

Stems: Simple or branched, slender, weak, often creeping with ascending branches and rooting at the nodes; to 20 inches tall.

Leaves: Basal and lower ones on slender stalks, upper sessile; blades pinnatifid, oblong in outline, ¹/₂ to 1¹/₂ inches long, divided into narrow, irregularly toothed or lobed segments usually reaching the midrib.

Flowers: Sepals 4, green, small; petals 4, exceeding the sepals; stamens 6; pistil 1, style short, thick. April through September.

Fruits: Very narrow, slightly curved to straight, can be spreading, ascending, or reflexed, and almost equal to the threadlike stalks.

Distribution: Wet to moist ground, especially in roadside ditches, low fields, along waterways. Common.

In Kentucky: (IP-mostly Inner Bluegrass), (ME-rare).

Also called creeping yellow watercress, this moisture-loving species is considered a noxious weed in many states because it has the ability to spread quickly by underground rhizomes. It has been introduced into nursery fields through contaminated roots and bulbs of lilies, daylilies, and hostas. Once established, it is difficult to eradicate, as a small fragment of root can produce an enormous number of new plants in one year.

Pale corydalis
Fumitory Family
Corydalis flavula (Raf.) DC.
Fumariaceae

Key features: Leaves finely divided into many narrow irregularly lobed segments; flowers pale yellow with saclike spur at the base and top petal crested.

Origin: Native.

Life form: Annual herb from a taproot.

Stems: One to several, erect to reclining, branching below, leafy, reddish green with whitish tinge; 6 to 10 inches tall.

Leaves: Alternate, blades dull green with whitish tinge, tips sharp-pointed or with a needlike point; lower long-stalked, upper sessile or short-stalked.

Flowers: In axillary clusters equaling or barely exceeding the leaves; sepals 2, falling early; petals 4, unequal, fused around 6 anthers, ¼ inch long; involucral bracts green. April through May.

Fruits: Capsule linear, about 1 inch long, drooping from a threadlike stalk.

Distribution: Moist to dry woods, rocky outcrops. Uncommon.

In Kentucky: AP, IP, (ME-rare).

The small yellow flowers on this spring wildflower often go unnoticed, blending in with the surrounding leaf litter.

The saclike spur at the base of the upper petal is responsible for the genus name, *Corydalis,* because it reminded early botanists of the crested lark (*korydalos* in Greek), a bird with a spur. Other common names are yellow-harlequin and yellow fumewort.

Yellow star-grass
Star-grass Family
Hypoxis hirsuta (L.) Coville
Hypoxidaceae

Key features: Plant from a corm; leaves basal, grasslike; flowers bright yellow, star-shaped, in umbels of 2 to 7.

Origin: Native.

Life form: Perennial herb.

Stems: Lacking except for flowering stalk 3 to 7 inches tall.

Leaves: Blades linear, margins entire, hairy; up to 12 inches long.

Flowers: Tepals 6, oblong to elliptic, spreading; stamens 6, anthers yellowish orange; pistil 1; ½ to ¾ inch wide. April through May.

Fruits: Capsule oval; seeds several, black, glossy with rows of warty projections.

Distribution: Dry open woods. Uncommon.

In Kentucky: AP, IP, ME.

The common name describes the starlike flowers and grasslike leaves. The underside of the tepals and stem are hairy—hence the species name, *hirsutus,* which is Latin for "hairy."

The Cherokee Indians used it to make a tea for strengthening the heart.

Spicebush
Laurel Family
Lindera benzoin (L.) Blume
Lauraceae

Key features: Aromatic understory deciduous shrub; leaves alternate, simple; fruits 1-seeded, turning bright shiny red in autumn.

Origin: Native.

Life form: Shrub, multi-stemmed with loose, open branches; to 12 feet tall.

Leaves: Blades elliptic, light green above, paler below, smooth, margins entire, bases wedge-shaped, tips short pointed to rounded; 3 to 5 inches long.

Flowers: Small, male and female on separate plants in few-flowered axillary clusters; sepals 6, yellow; petals absent; stamens 6; pistil 1. March through April.

Fruits: Drupe oval, green, turning red, ½ inch long; seed 1.

Distribution: Moist woods, woodland edges. Common in Iroquois; uncommon elsewhere.

In Kentucky: AP, IP, ME.

This graceful native shrub is one of the first to flower in early spring. The bright red drupes, on female plants only, ripen in the fall and are relished by birds. Larva of the spicebush swallowtail butterfly feed on the leaves.

Native Americans made a tea from all parts of the plant that was used to treat coughs, cramps, and measles. A bark tea was taken to expel worms.

The crushed leaves give off a delightful lemon scent.

Yellow fawn-lily

Lily Family
Erythronium americanum Ker Gawl.
Liliaceae

Key features: Plant from a bulb and stolons; leaves narrow, mottled brown, purple or green; flowers solitary, bright yellow, nodding at the end of a slender stalk.

Origin: Native.

Life form: Perennial herb.

Stems: Lacking except for flowering stalk 3 to 10 inches tall.

Leaves: Upright, single leaf when young, 2 when mature, tapering at the base into a clasping stalk; blades linear to elliptic, fleshy, margins entire; 3 to 7 inches long.

Flowers: Showy, tepals 6, yellow with reddish brown spots inside, becoming reflexed when mature; stamens 6, anthers yellow to brown; ovary 3-angled; 1½ inches wide. March through April.

Fruits: Capsule obovoid, 3-valved.

Distribution: Moist woods, slopes. Common in Cherokee; rare in Seneca; locally common in Iroquois.

In Kentucky: AP, IP, ME.

This beautiful wildflower is also called yellow adders-tongue, dog tooth violet, yellow snowdrop, and yellow trout-lily. Often forming large colonies in Cherokee and Iroquois, populations of nonflowering, stolon-producing plants with one leaf outnumber the flowering ones with two leaves.

Iroquois women made a root poultice to draw out splinters and reduce swellings and ulcers.

The generic name is derived from the Greek *erythros* "red" and refers to the red-flowered European species.

Invasive plant
Cher, Sen, Iroq, Shaw, Chick

Common yellow wood-sorrel

Wood-sorrel Family
Oxalis stricta L.
Oxalidaceae

Key features: Plant from extensive rhizomes, often mat-forming; leaves alternate, leaflets in groups of 3, clover-like, notched at the tip; flowers yellow.

Origin: Native.

Life form: Perennial herb.

Stems: Erect or ascending, freely branched at the base, weak, hairy, often rooting at the lower nodes; to 20 inches tall.

Leaves: On long slender stalks; blades heart-shaped, pale to grayish green, margin entire, smooth to densely hairy; ½ to 1 inch wide.

Flowers: Solitary or in few-flowered clusters at the tips of the stems; sepals 5, narrow, hairy; petals 5, spreading; stamens 10; styles 5; ½ inch wide. April through May and sporadically throughout the year.

Fruits: Capsule cylindrical, 5-ridged, short-beaked, about ½ inch long, on stalks that bend backward when mature; seeds several, tiny, orange-brown, ridged.

Distribution: Disturbed ground, especially in turf, fields, cultivated beds, open woods, roadsides, thickets. Common weed.

In Kentucky: AP, IP, ME. Listed as a Lesser Threat by the Kentucky Exotic Pest Plant Council.

Similar species: Illinois wood-sorrel (*Oxalis illinoensis* Schwegm.) is uncommon to rare and scattered throughout its range in northwestern Kentucky, Indiana, and western Tennessee; it is endangered in Illinois. Growing in moist woods and mossy banks in Cherokee, this plant is easily overlooked because it looks like the above weedy member. It has **slightly larger yellow flowers,** with green heart-shaped leaflets and **white,**

Yellow/Orange Flowers—Spring **149**

spindle-shaped underground tubers connected by slender rhizomes. Rare. April through May. In Kentucky: (IP-rare), (ME-scattered).

Common yellow wood-sorrel is also called pickles because the young green fruits and leaves taste sour. Another name, salt weed, comes from the Native Americans, who chewed the leaves on long walks to relieve thirst from loss of salt through perspiration.

Wood-sorrels are used as a folk remedy in Europe, Asia, and North America for the treatment of cancer. They also yield an orange dye.

Celandine poppy

Poppy Family
Stylophorum diphyllum (Michx.) Nutt.
Papaveraceae

Key features: Plant poisonous, with yellow sap; leaves shallowly or deeply pinnately divided into 5-to-7-lobed or toothed segments; flowers showy, bright yellow.

Origin: Native.

Life form: Perennial herb from rhizomes.

Stems: Erect or ascending, hollow, slightly hairy, leafless below; to 1½ feet tall.

Leaves: Basal, long-stalked; blades divided, segments oblong to obovate, slightly winged at base, dark green above, gray-blue below; stem leaves similar, smaller; 3 to 8 inches long. April through June.

Flowers: Solitary or few at top of stems; sepals 2, hairy, falling early; petals 4; stamens numerous, anthers yellow; pistil 1; 1½ to 2½ inches wide. April through June.

Fruits: Capsules elliptical, divided into 4 segments, densely hairy, 1 inch long; seeds numerous, shiny, dark brown.

Distribution: Moist wooded slopes. Common in Cherokee: uncommon in Iroquois.

In Kentucky: AP, IP.

This beautiful spring wildflower is also known as wood poppy and yellow poppy. The plant has a yellow sap that is poisonous.

Lesser celandine

Buttercup Family
Ranunculus ficaria L. subsp. *ficariiformis*
Ranunculaceae

Key features: Plant spreading rapidly from underground fingerlike tubers and bulblets below and bulbils in the leaf axils above; leaves dark glossy green; flowers bright yellow, solitary.

Origin: Eurasia.

Life form: Perennial herb.

Stems: Lacking except for the flowering stalk to 10 inches tall.

Leaves: Basal, long-stalked; blades heart-shaped to round, margins wavy to bluntly toothed, tip rounded; upper leaves few, smaller, with bulbils developing in the leaf axils; $1/2$ to 2 inches wide.

Flowers: Terminal, showy, on a smooth stalk; sepals 3, green, margin purple or white; petals 8 to 12, turning white with age, spreading; 1 inch wide. January through May.

Fruits: Achenes small, clustered on rounded heads.

Distribution: Disturbed ground, low moist to wet woods to dry uplands, waterways, limestone ledges. Invasive along Beargrass Creek; in pockets along the Ohio River.

In Kentucky: Collected in 4 counties, all along the Ohio River: Jefferson, Gallatin, Campbell, Lewis. Listed as a Severe Threat by the Kentucky Exotic Pest Plant Council.

Lesser celandine, also known as fig buttercup or spring messenger, was introduced into the United States in 1890 as an ornamental groundcover. Native to Eurasia, its greatest impact is on the native spring wildflowers that are unable to grow through the dense carpet of glossy green "slick" leaves. The plant spreads by means of bulblets, bulbils, and fingerlike tubers that break off from the parent plant and form new ones: these are easily transported by water currents during flooding events and heavy rains. By June, the above-ground plants die back and totally disappear from sight.

The word "celandine" comes from the Latin *chelidonia* and means "swallow." It was said that the flowers bloomed when the swallows returned in spring and faded when they left.

The scientific name *ficaria*—Latin for "fig"—describes the root tubers, which are said to resemble bunches of figs.

Hairy small-flowered crowfoot

Buttercup Family
Ranunculus micranthus (A. Gray) Nutt.
ex Torr. & A. Gray
Ranunculaceae

Key features: Stems hairy, weak, often forked above; leaves basal and alternate above; flowers tiny, yellow, turning whitish yellow.

Origin: Native.

Life form: Summer annual from fibrous roots.

Stems: Erect, slender, single- or multi-stemmed; 4 to 20 inches tall.

Leaves: Basal, long-stalked; blades round to broadly ovate, surfaces hairy, margins bluntly toothed, bases squared; upper ones reduced, clasping to short-stalked, divided into 3 to 5 lobes, linear to oblong, hairy, margins entire.

Flowers: Solitary to few, terminal, on long hairy stalks; sepals 5, ovate to elliptic, veined, spreading, becoming reflexed when mature; petals 5, veined, narrow, shorter than sepals; stamens many; pistils many. March through May.

Fruits: Achenes with tiny beak, grouped in compact round to cylindrical clusters.

Distribution: Disturbed wet or dry ground in open woods, waterways, fields, roadsides, thickets. Common.

In Kentucky: AP, IP, ME.

Similar species: Small-flowered crowfoot (*Ranunculus abortivus* L.) is a native summer annual growing to 20 inches tall. It looks exactly the same but has **smooth stems**. Cher, Sen, Iroq, Shaw, Chick. Disturbed sites as above. Common. March through May. In Kentucky: AP, IP, ME.

The generic name derives from the Latin *rana* "frog" and refers to the aquatic habits of some members. The species name, from the Latin *abortus,* means "miscarriage or incomplete" and alludes to the inconspicuous petals that give the impression of an incomplete flower.

Hooked buttercup

Buttercup Family
Ranunculus recurvatus Poir.
Ranunculaceae

Key features: Leaves basal below, alternate above, palmately divided into 3 to 5 lobes; flowers small, pale yellow, in terminal clusters; fruits beaked.

Origin: Native.

Life form: Perennial herb from fibrous roots.

Stems: Erect to ascending, single to branched above, light green, hairy; 1 to 2 feet tall.

Leaves: Basal on long hairy stalks; palmately divided, upper surface dull green, lower grayish, veins pale, margins sharply to bluntly toothed; stem leaves similar, smaller, uppermost stalkless; 1½ to 5 inches wide.

Flowers: Sepals 5, green, becoming reflexed; petals 5, shorter than sepals; stamens many, forming a ring around dense green pistils; ¼ inch wide. April through June.

Fruits: Achenes many, in rounded clusters, green turning black with age; beaks firm, distinctly hooked.

Distribution: Wet to moist open woods, waterways, roadside ditches. Common.

In Kentucky: AP, IP, ME.

The Cherokee Indians used the juice as a gargle for sore throats, and the Iroquois made a decoction from the roots that was taken to "kill the worms" in sore or hollow teeth.

Indian strawberry

Rose Family
Duchesnea indica (Andr.) Focke
Rosaceae

Key features: Plant trailing and rooting at the nodes; leaves compound, leaflets 3, coarsely toothed; flowers yellow, solitary in the leaf axils; fruits strawberry-like.

Origin: Asia.

Life form: Perennial herb from a crown with short rhizomes and stolons.

Stems: Many from the base, hairy; to 24 inches long.

Leaves: Basal and alternate; blades ovate to obovate, apex rounded; middle leaflet wedge-shaped at base, 2 lateral leaflets with unequal bases; to ¾ inch long.

Flowers: Sepals 5, green, triangular; petals 5, separate; stamens 20 to 25, anthers and filaments yellow; ovary purplish, pistils numerous; bracts below 5, spreading and alternating with the sepals; ½ to ¾ inch wide. April through August.

Fruits: Achenes clustered, round, fleshy, bright red.

Distribution: Disturbed ground in woods, turf, fields, cultivated beds, roadsides, waterways. Common to abundant.

In Kentucky: (AP-rare), IP, (ME-rare). Listed as a Lesser Threat by the Kentucky Exotic Pest Plant Council.

This ornamental plant has escaped from cultivation and is a common lawn weed that often misses the blades of a lawn mower due to its low-growing, mat-forming habit. It is often confused with wild strawberry (*Fragaria virginiana*), with white flowers and pleasant-tasting fruit, and with members in the genus *Potentilla* that have yellow flowers and five leaflets.

In Asia, a whole plant tea was traditionally made and used for laryngitis, coughs, and lung ailments, while a flower tea was used to stimulate blood circulation.

Running five-fingers

Rose Family
Potentilla canadensis L.
Rosaceae

Key features: Plant from stolons creeping along ground; leaflets 5, margins toothed above middle, entire below; flowers solitary, bright yellow, produced in the first node in leaf axil.

Origin: Native.

Life form: Perennial herb.

Stems: Silky hairy, erect or ascending, gradually elongating, often rooting at the nodes; to 20 inches long or more.

Leaves: Alternate, not well expanded during flowering time; palmately compound, blades obovate, middle leaflet largest, lower surface silky hairy, bases wedge-shaped, tips rounded; up to 1½ inches long.

Flowers: On threadlike, silky-hairy stalks; sepals deeply 5 cleft, interspersed with 5 sepal-like bracts; petals 5, rounded; stamens many; pistils many; about ½ inch wide. April through June.

Fruits: Achenes smooth, brown, in rounded heads.

Distribution: Open woods, fields, turf. Uncommon.

In Kentucky: AP, IP.

Similar species: Old field five-fingers (*Potentilla simplex* Michx.) is a coarser plant with hairy stems creeping to 24 inches long. Palmately compound, the **5 leaflets are narrowly oblanceolate with margins toothed** ¾ their length and well expanded at flowering time. The first **yellow flower is borne at the second node in the leaf axils** on the stem. Iroq. Open dry to moist woods, fields, roadsides. Prefers acid soil. Uncommon. April through June. In Kentucky: AP, IP, ME.

Old field five-fingers is also called decumbent five-fingers, which describes the low-creeping habit of the stems with five leaflets. It is mildly astringent, and early settlers used it as a gargle for loose teeth and spongy gums.

Spring avens

Rose Family
Geum vernum (Raf.) Torr. & A. Gray
Rosaceae

Key features: Leaves variable from round to heart-shaped, shallowly 3-to-5-lobed to deeply pinnately lobed; flowers few, yellow, in loose terminal clusters on long stalks.

Origin: Native.

Life form: Perennial herb from rhizomes.

Stems: Several, erect to arched-ascending, branched above, smooth to hairy; 12 to 20 inches tall.

Leaves: Basal ones long-stalked; blades variable, margins blunt to sharply lobed, hairs appressed; alternate middle leaves reduced upward, short-stalked, blades pinnately lobed, margins toothed; upper ones smaller, in 3s; leaflike stipules in pairs at base of large leaves.

Flowers: Sepals 5, ovate-triangular, recurved, margins hairy; petals 5, yellow, small, separate; stamens many, anthers yellow; pistils many; 1/4 inch wide. April through June.

Fruits: Heads round on a distinct stalk; achenes bristly, bearing persistent, 1-jointed bent or looped barbs.

Distribution: Moist woods, waterways, bottomlands, fields, disturbed ground. Common.

In Kentucky: AP, IP, ME.

This native wildflower is often found in compacted soil and is especially common in picnic grounds, where there is little competition from other plants.

It is reported that this more southern plant is now moving northward in response to warming winters and becoming more widespread, especially in disturbed sites.

Smooth yellow violet

Violet Family
Viola pubescens Aiton.
Violaceae

Key features: Stems 2 to 3 or more; leaves basal, 3 or more, long-stalked, upper leaves 2 to 4, short-stalked, heart-shaped; flowers yellow with purple veins near base.

Origin: Native.

Life form: Perennial from short caudex with fibrous roots.

Stems: Erect, hairy to smooth; 4 to 12 inches tall.

Leaves: Blades broadly heart-shaped to orbicular, margins bluntly to finely short-toothed, silvery white hairs on veins below or both on the veins below and in between; leaflike stipules lance-ovate, margins entire to finely toothed.

Flowers: Spring ones produced from upper leaf axils on stalks sparsely to densely hairy; sepals 5, narrow; petals 5, upper petals 2, lateral petals 2, bearded, lower purple-veined; ¾ inch wide. April through May.

Fruits: Capsule ovoid, smooth to woolly; seeds small, brown.

Distribution: Moist woods, especially in alluvial soils. Uncommon.

In Kentucky: AP, IP, ME.

Throughout antiquity, violets have been valued as a source of medicine and food. They have been likened to spinach and have been widely used as a potherb for centuries.

In Appalachia, they are sugared and used as a candy called "messages of love."

Largeflower bellwort

Bellwort Family
Uvularia grandiflora L.
Uvulariaceae

Key features: Stems piercing the leaves, blades hairy below; flowers with 6 yellow tepals, bellow-shaped, smooth within.

Origin: Native.

Life form: Perennial herb from rhizomes, often forming colonies.

Stems: Young plants usually single-stemmed, dividing into 2 to 3 when mature, upper portion nodding, light green with whitish tinge, smooth; to 18 inches tall.

Leaves: Alternate, blades oblong-ovate to lance-shaped, margins smooth, parallel-veined, upper surface green, smooth, below greenish white, tips tapering; 3 to 6 inches long. April through May.

Flowers: Terminal, dangling; tepals 6, narrow, twisted; stamens 6, anthers yellow; styles 3-forked; 1 to 2 inches long.

Fruits: Capsules 3-lobed; seeds several.

Distribution: Rich moist slopes, especially limestone. Rare.

In Kentucky: AP, IP, ME.

Similar species: Perfoliate bellwort (*Uvularia perfoliata* L.) is a native perennial herb that grows to 12 inches tall. The nodding terminal flowers with 6 yellow **tepals have minute granular yellow bumps inside.** The **stem pierces the leaf blade, which is smooth on the underside.** Iroq. Rich moist slopes. Rare. April through May. In Kentucky: AP, IP.

Linnaeus, the great Swedish botanist, named this genus *Uvularia* because the dangling petals resembled the uvula (flap of tissue hanging down from the back of the mouth).

Native Americans made a root tea that was used as a wash for rheumatic pains and mixed with fat as an ointment for sore muscles, tendons, backaches, and lung ailments.

Devil's beggar-ticks
Aster Family
Bidens frondosa L.
Asteraceae

Key features: Leaves opposite, pinnately compound; flower heads yellow, often without ray florets, green bracts 5 to 8.

Origin: Native.

Life form: Summer annual herb from a taproot.

Stems: Slender, sometimes branched above, green or reddish purple, mostly smooth; to 3 feet tall.

Leaves: Leaflets 3 to 5, narrow, terminal leaflet largest, margins coarsely toothed, tips long pointed; upper leaves reduced, narrow, toothed; 1 to 6 inches long.

Flowers: In terminal heads, long-stalked; inner disk florets orange, tiny; outer ray florets yellow, often absent or rudimentary; involucral bracts narrow, usually longer than disk florets, margins hairy. June through October.

Fruits: Achenes flat, black with 1 prominent vein, awns 2, with backward-pointing barbs.

Distribution: Moist or wet disturbed ground, especially in roadside ditches, waterways, thickets, fields. Common.

In Kentucky: AP, IP, ME.

Similar species: Spanish-needles (*Bidens bipinnata* L.) is a summer annual with erect stems to 4 feet tall. The **leaves mostly 2 to 3 times pinnately dissected** with irregular lobes. The flower heads are yellow with **7 to 10 outer, narrow, green bracts.** Achenes narrow, 4-sided with **3 to 4** awns with backward-pointing barbs. Cher, Sen, Iroq, Shaw, Chick. Moist or wet disturbed ground. Common. June through October. In Kentucky: AP, IP, ME.

The common name, Spanish-needles, refers to the three to four barbed awns—or "needles"—on the black seeds, which stick to the clothes of passersby and easily travel away from the parent plant. The genus includes the tickseeds, bur-marigolds, and beggar-ticks.

Divaricate sunflower

Aster Family
Helianthus divaricatus L.
Asteraceae

Key features: Leaves opposite, narrow, spreading horizontally in remote pairs; flower heads with outer yellow ray florets surrounding a yellow center.

Origin: Native.

Life form: Perennial herb from long, creeping rhizomes.

Stems: Erect, smooth to slightly hairy, simple or branched above, often with a whitish bloom; 1½ to 4 feet tall.

Leaves: Blades lanceolate to ovate, surfaces rough covered with tiny white dots, 3-nerved, margins entire to toothed, bases rounded, sessile to short-stalked, tips sharp-pointed to tapering; 1½ to 6 inches long.

Flowers: Terminal heads solitary to several; inner disk florets many, small, tubular; outer ray florets narrow, 8 to 15; involucral bracts narrow, overlapping, becoming recurved, tips short-pointed, margins white-hairy; 1 to 2 inches wide. July through September.

Fruits: Achenes small, dark brown.

Distribution: Open dry woods, woodland edges, roadsides. Common.

In Kentucky: AP, IP, ME.

The genus *Helianthus* contains many species that are the "true" sunflowers. The name comes from the Greek *helios* "sun" and *anthos* "flower."

Farmers and garden enthusiasts who grow sunflowers tell stories about a peculiar behavior called "solar tracking," where the flower follows the movement of the sun. As soon as the sun rises, seedlings, buds, and leaves on young plants face east and track the sun across the sky, turning westward by dusk. By dawn the next morning, they have returned eastward and are ready to start the journey again.

This common wildflower is also called woodland sunflower, which describes its preferred habitat.

Ashy sunflower

Aster Family
Helianthus mollis Lam.
Asteraceae

Key features: Stems stout, grayish green, hairy; leaves opposite, sessile, bluish gray; flower heads terminal, showy, with a yellowish green center and outer yellow ray florets.

Origin: Native.

Life form: Perennial herb from rhizomes, often forming colonies.

Stems: Erect, 1 to several from the base, densely hairy; to 3 feet tall.

Leaves: Blades ovate, stiff, 3-nerved, densely hairy, margins finely toothed, bases rounded to slightly heart-shaped, tips pointed; to 5 inches long.

Flowers: Inner disk florets many, tubular; outer ray florets linear-lanceolate, 15 to 30; involucral bracts narrow, purple-tipped, loosely ascending, densely hairy; 1 to 2½ inches wide. July through September.

Fruits: Achenes 3-angled, brown, densely hairy.

Distribution: Uncommon at Cherokee; common at Summit Field.

In Kentucky: (IP-mostly in Shawnee Hills), ME.

This attractive sunflower is one of the most widespread species of upland prairies in the tallgrass region from northern Illinois southward. In Kentucky, it is found growing in the western part of the state in prairie patches and fields but is not common. It often forms dense colonies from creeping rhizomes and produces allelopathic chemicals that inhibit the growth of other plants nearby. Large showy patches of ashy sunflower may be seen in their glory in late summer at Summit Field.

Beaked hawkweed

Aster Family
Hieracium gronovii L.
Asteraceae

Key features: Leaves mostly basal, one or two on lower portion, reduced upward and sessile, leafless above; flower heads yellow with a minute bract at base of flower stalk.

Origin: Native.

Life form: Perennial herb from fibrous roots and rhizomes.

Stems: Mostly solitary, erect, densely hairy below, hairs star-shaped, appressed or long-spreading; 1 to 3 feet tall.

Leaves: Blades oblanceolate to elliptic, upper surface hairy, bases black-dotted, midvein below densely hairy, margins entire, hairy, tips short pointed or rounded; to 5 inches long. June through September.

Flowers: Heads in a cylindrical, elongated open cluster with dark glandular hairs; ray florets yellow, tip 5-toothed, style yellow, 2-forked; involucral bracts in 2 series, short outside, longer inside, with dark glandular hairs. June through September.

Fruits: Achenes black to burgundy, distinctly narrowed toward top, ribbed; tufts of hairs tawny.

Distribution: Dry open woods, especially along eroded trails. Uncommon.

In Kentucky: AP, IP, ME.

The scientific name comes from *hierakion*—from the Greek *hierax* "hawk"—and refers to an ancient superstition that hawks feed on this plant to strengthen their eyesight. Herb doctors used this plant in eye lotions, and the fresh juice from the leaves was said to cure warts.

Orange dwarf-dandelion
Aster Family
Krigia biflora (Walter) S.F. Blake
Asteraceae

Key features: Stems bluish green, forking above; both basal and stem leaves present; flower heads several, terminal, orange, dandelion-like.

Origin: Native.

Life form: Perennial herb from fibrous roots.

Stems: Nearly leafless, smooth, with milky sap; 1 to 2 feet tall.

Leaves: Mostly basal, blades oblanceolate to broadly elliptic, tapering at base, bluish green above, paler below, margins entire, toothed, lobed or cleft; upper ones few, reduced, clasping; 1 to 7 inches long.

Flowers: Ray florets notched at the tip; involucral bracts green, narrow, margins pale, becoming reflexed with age; 1 inch wide. May through October.

Fruits: Achenes brown, slightly ribbed, topped with a tuft of whitish tan unequal bristles.

Distribution: Open woods, roadsides, fields. Common.

In Kentucky: AP, (IP-rare in Inner Bluegrass), ME.

Similar species: Colonial dwarf-dandelion [*Krigia dandelion* (L.) Nutt.] is a colonial species with underground potato-like tubers. **Basal leaves,** up to 8 inches long, are lanceolate with entire, toothed, or lobed margins. **The leafless stems bear a solitary yellow flower head** about 1 inch wide. Subtending the heads are 9 to 18 green, involucral bracts that become reflexed with age. Iroq. Moist woods. Rare. April through June. In Kentucky: (IP-rare, absent from the Bluegrass), ME.

The genus honors David Krieg, a German physican, who was one of the first to collect plants in Maryland in the late seventeenth and early eighteenth centuries.

The Menominee Indians used the stems to make a sound that mimicked a fawn in distress and lured the doe to the hunter.

Colonial dwarf-dandelion is also called potato dandelion and refers to the underground potato-like tubers.

Prickly lettuce

Aster Family
Lactuca serriola L.
Asteraceae
(*syn=Lactuca scariola* L.)

Key features: Stems contain white latex; leaves arrowhead-shaped, prickly, clasping the stem; flower heads yellow, dandelion-like, borne in loose terminal clusters.

Origin: Europe.

Life form: Winter annual or biennial herb from a taproot.

Stems: Erect, light green or yellowish, branching above; to 7 feet tall.

Leaves: Alternate; lower ones deeply to shallowly lobed with one to several lobes on each side; the upper ones smaller, narrow; all slightly curved backward, margins and midvein below prickly; 3 to 12 inches long.

Flowers: Many, small; outer ray florets yellow, 13 to 24 each with a dark blue stripe on lower side; involucral bracts green, in 3 to 4 series, longer inward. July through September.

Fruits: Achenes small, black or gray, flattened with 5 to 7 ridges, spiny above, beak long, crowned at tip with tufts of white hairs.

Distribution: Disturbed ground, fields, roadsides, thickets, woodland edges, cultivated beds. Common.

In Kentucky: AP, IP, ME.

Prickly lettuce contains more Vitamin A than spinach and a large quantity of Vitamin C. The young tender leaves are good in salads. The cultivated lettuce of today, which has its origin in the Mediterranean region and Near East, is believed to be an ancient cultigen derived from prickly lettuce.

Another common name is wild opium and refers to the compound lactucarium—or "lettuce opium"—found in the milky sap, which is mildly sedating. It can also cause dermatitis in people with sensitive skin.

Pineapple-weed

Aster Family
Matricaria discoidea DC.
Asteraceae
[*syn=Matricaria matricarioides* (Less.)
Porter]

Key features: Plants smell like pine-apple; leaves fernlike; flower heads yellowish green, cone-shaped.

Origin: Adventive from the Pacific coast and spreading eastward.

Life form: Summer annual from a taproot.

Stems: Erect, branched to bushy; 2 to 16 inches tall.

Leaves: Alternate, numerous, light green, very finely divided.

Flowers: Heads short-stalked, solitary, terminal; disk florets only, yellowish green; involucral bracts overlapping, green with a papery margin. May through August.

Fruits: Achenes small, yellowish gray, obovate to oblong, 3-to-5-ribbed, beaked.

Distribution: Disturbed ground, especially along gravelly roadsides. Uncommon; locally common in Seneca.

In Kentucky: (AP-rare), IP, (ME-rare).

Related to the mayweeds and chamomiles, this plant imparts a fragrance of pineapple when the leaves or flowers are crushed or bruised.

The Blackfoot Indians gathered and dried the small, cone-shaped flower heads and used them to make perfume and insect repellent. Traditionally, an infusion made with other plants was used externally for sores and as a wash to help relieve the itch from poison-ivy and other rashes.

Pineapple-weed is often found growing along roadsides. The small seeds stick to the tires of cars and bikes as well as shoes, which helps spread the plant.

Yellow coneflower

Aster Family
Ratibida pinnata (Vent.) Barnhart
Asteraceae

Key features: Leaves pinnately divided below, upper narrow, smaller; tiny inner disk florets borne on a gray-green to dark green conical center surrounded by yellow, drooping ray florets.

Origin: Native.

Life form: Perennial herb from rhizomes.

Stems: Multiple from base, erect, slender, branching above; 2 to 5 feet tall.

Leaves: Alternate, lower long-stalked, blades divided into 3 to 7 (11) lobes, these sometimes divided again, margins coarsely toothed to entire; stem leaves short-stalked to sessile, blades lanceolate; to 7 inches long.

Flowers: Heads solitary to several on long stalks at the tips of the branches; inner disk florets tubular, 5-lobed; stamens 5, anthers purplish brown, stigma dark, style 2-forked; outer yellow ray florets 4 to 10; involucral bracts green, spreading, reflexed; 3 inches wide. June through August.

Fruits: Achenes tiny, white, smooth and flattish.

Distribution: Woodland edges, thickets, fields. Common at Cherokee; uncommon Iroquois.

In Kentucky: IP, (ME-rare).

Also known as gray-headed coneflower because the oblong central cone is grayish green at first and later turns brown. The plant is said to emit an odor of anise when crushed.

Cutleaf black-eyed Susan

Aster Family
Rudbeckia laciniata L.
Asteraceae

Key features: Leaves alternate, large, divided into 3 to 7 lobes; flower heads greenish yellow, on a rounded to cone-shaped head, showy.

Origin: Native.

Life form: Perennial herb from fibrous roots.

Stems: Erect, smooth, light green often with a white tinge; 3 to 6 feet tall.

Leaves: Lower to 12 inches long, stalked, upper shorter; blades lobed, margins coarsely toothed or narrowly incised, upper surface dark green, smooth, slightly hairy below, bases wedge-shaped, tips tapering.

Flowers: Solitary or few-flowered, terminal, on long stalks; inner disk florets many, small, tubular, greenish yellow; outer ray florets yellow, 6 to 10, drooping, tip notched; involucral bracts green, unequal, narrow, often reflexed, in 1 to 2 series; 2 to 4 inches wide. July through September.

Fruits: Achenes flattened, 4-angled, unequal at base, pinkish when young, turning black when mature.

Distribution: Moist soil, thickets, waterways. Uncommon.

In Kentucky: AP, IP, ME.

This native prairie wildflower has become a popular garden plant and is also called cutleaf coneflower, green-headed coneflower, and goldenglow. The species name, *laciniata,* means "torn" and refers to the incised, torn-looking leaves.

Native Americans gathered and cooked the young shoots and leaves in the spring and ate them for "good health." However, the plant is toxic if ingested in large quantities.

A dressing made from the blossoms was used to treat burns.

Three-lobed coneflower

Aster Family
Rudbeckia triloba L.
Asteraceae

Key features: Leaves usually 3-lobed below, narrow above; flower heads showy, the purplish black cone-shaped center is surrounded by yellow ray florets; bracts bristle-tipped.

Origin: Native.

Life form: Biennial or short-lived perennial with fibrous roots.

Stems: Erect, bushy, branched, reddish green, hairs white; 2 to 3 feet tall.

Leaves: Lower blades stalked, narrowly winged at base, surface rough, margins coarsely toothed to entire; middle and upper blades ovate to lanceolate, sessile; to 4 inches long.

Flowers: Heads solitary, at the top of stem branches, about 2 inches wide; inner disk florets many, tubular, 5-lobed; outer ray florets, 6 to 12, notched at tip; style 2-forked; involucral bracts green, usually 8, recurved, in a single series. June through October.

Fruits: Achenes small, 4-angled, white, becoming black at maturity.

Distribution: Woodland edges, fields, roadsides, waterways. Common in Seneca; uncommon in Cherokee.

In Kentucky: (AP-rare), IP, (ME-rare).

Similar species: Black-eyed Susan (*Rudbeckia hirta* L.) is a native biennial or short-lived perennial herb to 3 feet tall. The long-stalked **basal leaves are elliptic to lanceolate; the middle and upper leaves are smaller, sessile and narrow.** The rough upper and lower surfaces are densely hairy with toothed margins. Solitary, terminal flowers have a purplish brown cone-shaped center surrounded by yellow outer ray florets. The involucral **bracts are sharp-pointed at tip, hairy.** Cher, Iroq. In thickets and fields where it has escaped mowing. Uncommon. June through October. In Kentucky: AP, IP, ME.

Three-lobed coneflower is a beautiful native wildflower that is drought, pest, and heat resistant and often sold in the nursery trade because of these qualities. The showy flowers can last up to three months and attract butterflies and other insects.

Other common names are thin-leaved coneflower and brown-eyed Susan. The latter name refers to the dark brown center or "eye" in the flower.

The genus was named by Carl Linnaeus in honor of his teacher at the University of Uppsala, Professor Olaf Rudbeck the younger, and his father, both of whom were botanists.

Cup-plant
Aster Family
Silphium perfoliatum L.
Asteraceae

Key features: Plant tall; leaves opposite, joined around the stem forming a cup; flower heads yellow, numerous.

Origin: Native.

Life form: Perennial herb from rhizomes.

Stems: Smooth, square, leafy; to 8 feet tall.

Leaves: Blades triangular to broadly ovate, midvein white, margins coarsely toothed, rough above, rough to short-hairy below; to 18 inches long.

Flowers: Inner disk florets small, tubular, sterile; outer ray florets yellow, narrow, veins pale green; involucral bracts ovate to elliptic, pale green, margins short-hairy; 2 to 4 inches wide. July through September.

Fruits: Achenes obovate, winged, with deep sinus at top.

Distribution: Moist areas along waterways, woodland edges, thickets. Uncommon.

In Kentucky: (AP-rare), IP, ME.

This stately native plant has a resinous juice, and the genus name is from the Greek word *silphion,* the ancient name of a North African resinous plant. The large, joined cuplike leaves can hold water.

Zigzag goldenrod

Aster Family
Solidago flexicaulis L.
Asteraceae

Key features: Stems zigzagged; leaves alternate, abruptly contracted into a winged stalk; flower heads golden yellow, in terminal and axillary clusters.

Origin: Native.

Life form: Perennial herb from long rhizomes.

Stems: Angled, grooved, smooth below the inflorescence; 1 to 4 feet tall.

Leaves: Blades ovate to elliptic, margins sharply toothed, lower surface with hairs on midrib and main veins, tips pointed; 3 to 6 inches long.

Flowers: Inner disk florets golden yellow, 5 to 9; ray florets yellow, 3 to 4; involucral bracts in 2 series, outer blunt-tipped, inner broadly rounded. July through October.

Fruis: Achenes short-hairy.

Distribution: Moist woods, moist limestone outcrops. Uncommon.

In Kentucky: AP, IP, (ME-rare).

This goldenrod is a treat to find and is especially beautiful when growing on limestone cliffs at Cherokee. The zigzagged stem makes identification easy.

Native Americans made a compound of leaves that was dried and powered and inserted in the nostrils to stop a nosebleed or headache. Chewed roots were an aid for a sore throat.

Key features: Stems hairy; leaves basal and alternate above; flower heads deep yellow, in small clusters crowded along a narrow, spikelike inflorescence to 10 inches long.

Origin: Native.

Life form: Perennial herb from a rhizome.

Stems: Erect, rough, ridged; 1 to 3 feet tall.

Leaves: Basal and lowermost with long winged stalks, blades ovate to elliptic, hairy, margins smooth or shallowly toothed; middle to upper blades similar, reduced, stalked to sessile; to 8 inches long.

Flowers: Inner disk florets yellow, 7 to 16, tubular; outer ray florets deep yellow, 7 to 14; involucral bracts narrowly rounded, yellowish, green tip ill-defined; ¼ inch wide. July through October.

Fruits: Achenes smooth.

Distribution: Dry to moist open woods. Common.

In Kentucky: AP, IP, ME.

Similar species: Silver-rod (*Solidago bicolor* L.) grows to 3 feet tall and has densely white hairy stems and leaves. The **disk and ray florets are white, occasionally pale yellow,** with a distinct green line and tip on the involucral bracts. These are crowded in a **dense spike to 15 inches long.** Iroq. Dry open upland woods. Uncommon. July through October. In Kentucky: AP, IP.

Of the many species of goldenrod found growing in Kentucky, silver-rod is distinct because it has white flowers instead of the characteristic bright yellow ones.

Hairy goldenrod

Aster Family
Solidago hispida Muhl. ex Willd
Asteraceae

Prickly sow-thistle

Aster Family
Sonchus asper (L.) Hill
Asteraceae

Key features: Plant with milky latex; leaves alternate, prickly, with clasping, heart-shaped bases; flower heads pale yellow, dandelion-like, in open clusters of 1 to 5.

Origin: Europe.

Life form: Summer annual herb from a taproot.

Stems: Erect, hollow, branched or unbranched, often reddish, leafy; to 5 feet tall.

Leaves: Blades obovate to oblanceolate, margins entire to many-lobed, tip lobe triangular; middle and upper leaves progressively reduced upward; 3 to 12 inches long.

Flowers: Ray florets only, tiny; involucral bracts with inner row longer than outer. June through October.

Fruits: Achenes tiny, light brown, ribbed, margins winged, tufts of white hairs soon separating.

Distribution: Disturbed ground along roadsides, fields, limestone ledges, thickets. Uncommon.

In Kentucky: AP, IP, ME.

Another common name is rough sowthistle, which describes the texture of the plant.

The generic name, *Sonchus,* is Latin for "hollow" and refers to the hollow stems, while *asper* means "rough." This weedy species invades disturbed sites throughout the world. It is capable of producing up to 26,000 seeds, which are viable for up to eight years.

In parts of Europe, the young leaves are used as a potherb or for salad and contain nearly three percent protein and are rich in calcium.

Species within the genus *Sonchus* have a milky juice, and the latex was thought to induce lactation in animals.

Orange touch-me-not

Touch-me-not Family
Impatiens capensis Meerb.
Balsaminaceae

Key features: Stems hollow, succulent, with watery juice; leaves alternate, stalked; flowers bright orange with red spots and a forward-curving spur.

Origin: Native.

Life form: Summer annual herb from a taproot.

Stems: Single or branched above, pale green to reddish green with whitish tinge; to 5 feet tall.

Leaves: Blades ovate to elliptic, dull green above, paler below, margins broadly toothed, bases rounded, tips with short point; 1 to 5 inches long.

Flowers: Few, dangling on slender stalks from upper leaf axils; sepals 3, petal-like, upper 2 small, green, 1 spurred; petals 3, 1 above, 2 laterals; stamens 5, anthers united into a cap; pistil 1; 1 inch long. May through September.

Fruits: Capsule fleshy, green, 5-valved, opening explosively expelling the many black seeds.

Distribution: Moist open woods, waterways, meadows, roadside ditches, thickets, pond margins. Common.

In Kentucky: AP, IP, ME.

Similar species: Yellow touch-me-not (*Impatiens pallida* Nutt.) has **yellow flowers with a spur that spreads slightly outward at a right angle.** The leaves are similar as above. Flowering at the same time, this species is often intermixed with the orange touch-me-not. Cher, Sen, Iroq, Shaw, Chick. Moist open woods, waterways, roadside ditches, thickets, pond margins. Uncommon. May through September. In Kentucky: AP, IP, (ME-rare).

The common names jewel weed and lady's-earrings refer to the dangling flowers, which are said to resemble jewels.

The plant has a silvery shine to it when a water droplet lands on the leaf surface or when held under water, and they are also called silverweeds. The generic name, *Impatiens,* is Latin for "impatient" and describes the mature seed pods that burst open when touched, scattering the seeds.

The hollow, succulent stems and spur sac contain a watery juice that is used to sooth the itch from poison-ivy and nettle stings. The Cherokee Indians used the juice from the blossoms to treat athlete's foot. Today, it is known that the juice does contain chemicals that act as fungicides.

Trumpet creeper

Bignonia Family
Campsis radicans (L.) Seem. ex Bureau
Bignoniaceae

Key features: A woody vine; leaves opposite, pinnate; flowers showy, orange-red, trumpet-shaped; fruits cigar-shaped.

Origin: Native.

Life form: Deciduous perennial vine from a taproot and extensive rhizomes.

Stems: High-climbing or trailing, several from base, smooth; to 30 feet or more.

Leaves: Short leaf stalk slightly winged; leaflets 5 to 11, blades ovate, bright green, veins whitish, margins coarsely toothed; tips short-pointed to long-pointed; 1/2 to 3 inches long.

Flowers: In terminal clusters; sepals 5-lobed, reddish tan; petals orange-red, funnel-form with 5 spreading lobes at tip; stamens 5, anthers tan; stigma flattened; 2 to 3 inches long. July through September.

Fruits: Capsules, woody, slender, 4 to 8 inches long, splitting open along 2 seams; seeds flattened, winged.

Distribution: Disturbed ground in open woods, thickets, roadsides; often climbing trees. Uncommon.

In Kentucky: AP, IP, ME.

This plant is also called cow-itch. People drinking milk from cows feeding on the vine may develop itchy skin. It is also said that when cows eat the plant an inflammation occurs on the udders and that those milking the cows develop rashes. Some people develop a rash just from handling the leaves or flowers.

Trumpet creeper climbs with the aid of aerial roots along its stem. Once it reaches a certain height, it produces horizontal branches that grow away from the support. Other names that describe this trailing and climbing habit of this aggressive plant are devil's shoelaces and hell vine.

Common St. John's-wort

St. John's-wort Family
Hypericum perforatum L.
Clusiaceae

Key features: Stems ridged below leaves: leaves opposite with translucent dots lining edge and underside of blade; flowers numerous, yellow-orange with black dots lining the margin.

Origin: Europe.

Life form: Perennial herb from rhizomes and basal offshoots.

Stems: Erect, branched above, light green; up to 2 feet tall.

Leaves: Blades linear-oblong to elliptic, clasping, margins entire; about 1 inch long.

Flowers: In open, flat-topped clusters; sepals 5, green, narrow, tips sharp; petals 5; stamens numerous, showy; pistil 1, styles 3; ¾ inch wide. July through August.

Fruits: Capsule oblong; seeds tiny, numerous, oblong, dark brown, mottled.

Distribution: Summit Field. Uncommon.

In Kentucky: (AP-rare), IP, ME. Listed as a Moderate Threat by the Kentucky Exotic Pest Plant Council.

Similar species: Spotted St. John's-wort (*Hypericum punctatum* Lam.) is a native perennial herb growing to 2½ feet tall. The **stems are rounded below** the opposite, short-stalked to sessile leaves. The blades are oblong, parallel-veined, and have **translucent dots on the underside**. Flowers are **small** with 5 **golden yellow, black-dotted petals** and numerous stamens. Cher, Iroq. Dry to moist open woods, especially along trails. Uncommon. June through August. In Kentucky: AP, IP, ME.

Many superstitions are associated with Common St. John's-wort. Named after John the Baptist, the plant is believed to show its black dots on the 29th of August, the day the prophet was beheaded.

Introduced as a garden ornamental, this species has escaped and is a weed in pastures, roadsides, fields, and meadows. It is poisonous to livestock who graze on this plant due to the photosensitive reaction and gastrointestinal distress. However, today, it is commercially produced and available in health food stores as an herbal remedy for calming nerves.

Partridge-pea

Legume Family
Chamaecrista fasciculata (Michx.) Greene
Fabaceae
(*syn=Cassia fasciculata* Michx.)

Key features: Leaves alternate, pinnately compound; flowers bright yellow, stamens yellow and reddish brown, unequal.

Origin: Native.

Life form: Summer annual herb from a taproot.

Stems: Erect or ascending, slender, hairy, light green becoming reddish brown; 2 to 3 feet tall.

Leaves: Leaflets 5 to 17, blades linear-oblong, tips blunt to abruptly pointed, ¾ inch long; leaflike stipules at base of leaf stalk narrow, triangular, with a dark round gland.

Flowers: Axillary, on slender stalks; sepals 5, narrow; petals 5, irregularly shaped, lowest petal largest, upper 4 with reddish spot at base; stamens 10; 1 inch wide. June through October.

Fruits: Pods green, flat, narrow, hairy when young, maturing brown and smooth, opening along 2 sutures; seeds several, pitted, in rows.

Distribution: Summit Field. Abundant.

In Kentucky: AP, IP, ME.

Partridge-pea is considered a pioneer species and often forms dense stands in disturbed, infertile ground where other plants cannot grow.

The showy yellow flowers attract several kinds of short-tongued bees and a variety of flies and ants because of the nectar gland at the base of the leaf stalk. It is also an important honey plant in some areas of the southern United States.

It is also called large-flowered senstitive-pea because the leaflets are sensitive when touched and will fold up.

Black medic

Legume Family
Medicago lupulina L.
Fabaceae

Key features: Stems spreading; leaves alternate, divided into 3 wedge-shaped leaflets; flowers tiny, yellow, pealike; fruit pods coiled.

Origin: Eurasia.

Life form: Winter annual or biennial from a taproot.

Stems: Ascending or trailing, pale green, ridged, hairy, branched at the base; to 3 feet long.

Leaves: On threadlike stalks, about 1 inch long; leaflets 3, terminal leaflet longer-stalked, margins toothed on upper half; leaflike stipules at base of leaf stalk ovate-lanceolate, tips long-pointed.

Flowers: In dense, rounded clusters on long stalks exceeding the leaves; sepals 5, green, narrow, long-pointed; petals 5; stamens 10. May through October.

Fruits: Pod coiled, 1-seeded, kidney-shaped, turning black at maturity; seeds yellowish green, oval to kidney-shaped, tiny with longitudinal veins.

Distribution: Disturbed ground in turf, cultivated beds, roadsides, fields. Common weed.

In Kentucky: AP, IP, ME. Listed as a Significant Threat by the Kentucky Exotic Pest Plant Council.

Black medic was introduced into North America as a forage crop. It has become a valuable addition to pasturage in the higher mountains of the western United States. Because of its superiority and adaptability to various climates and soils, the plant is also called nonesuch, meaning that no other plant is like it. However, it has spread, often as a contaminant in uncleaned alfalfa and clover seed, and is now a troublesome weed in turf, gardens, and disturbed ground throughout its range.

In Ireland, this species is often recognized as the "true shamrock."

Northern wild senna

Legume Family
Senna hebecarpa (Fernald) H.S. Irwin &
Barneby
Fabaceae

Key features: Leaves pinnately
compound; nectar gland at leaf base
club-shaped, short-stalked; flowers yellow
with black anthers.

Origin: Native.

Life form: Perennial herb from a taproot
and thick rhizomes.

Stems: Stout, erect, light green, smooth
to hairy above; 3 to 5 feet tall.

Leaves: Alternate, leaflets 10 to 20,
blades oblong to elliptic, margins entire,
tips rounded to sharp tipped; up to 2½
inches long.

Flowers: In several many-flowered
axillary clusters; sepals 5, unequal; petals
5 with 3 upper, 2 lower, unequal;
stamens 10, anthers black; pistil 1,
densely white hairy; ¾ inch wide. July
though August.

Fruits: Seed pods dangling, segments
10 to 18, nearly squared, hairs white,
loosely spreading; seeds flat, nearly as
long as wide.

Distribution: Moist open ground,
thickets. Uncommon.

In Kentucky: (AP-rare), IP.

This native wildflower has become a
popular garden plant and is capable of
forming large colonies via rhizomes and
seed production. It prefers rich alluvial
soils and is only found in several counties
in the Inner Bluegrass.

The glands on the leaf stalk secrete
sugary nectar that attracts ants and other
insects. The shape of these glands is also
an important identification characteristic.

Pencil-flower

Legume Family
Stylosanthes biflora (L.) BSP
Fabaceae

Key features: Stems wiry; leaflets 3, bristle-tipped, terminal leaflet stalked, slightly larger than lateral ones; flowers pealike, yellow.

Origin: Native.

Life form: Perennial herb from a taproot.

Stems: Branched from the base, erect or ascending; to 12 inches tall.

Leaves: Alternate, stalked; blades lanceolate to oblong, dark green above, paler below, midvein distinct, margins slightly bristly; to 2 inches long; stipules leaflike, surrounding the stem.

Flowers: Small, 2 to 6 in short leafy spikes in upper leaf axils; sepals green, unequally 4 to 5 cleft; petals 5, upper 1 largest, rounded, side wings 2, lower petal curved upward, about equaling the wings; ¼ to ½ inch long. July through September.

Fruits: Pods jointed, flattened, persistent style hooked.

Distribution: Open rocky woods, fields, turf. Common.

In Kentucky: AP, IP, ME.

The generic name is from the Greek *stylos* "column" and *anthos* "flower" and refers to the stalk-like calyx tube.

The Cherokee Indians used the root to help remedy female complaints.

St. Andrew's cross

Mangosteen Family
Hypericum stragulum W.P. Adams &
N. Robson
Clusiaceae
[*syn=Ascyrum hypericoides* L. *var.
multicaule* (Michx.) Fernald]

Key features: Low-growing woody
shrub; leaves opposite, pale bluish green
with minute black dots above and below;
flowers yellow, the petals forming an "X."

Origin: Native

Habitat: Woody shrub from slender
rhizomes.

Stems: Reddish brown, smooth,
mat-forming; to 9 inches tall.

Leaves: Blade obovate, midvein distinct,
margins entire to slightly wavy, narrowed
at the base, tips rounded; ¼ to 1 inch long.

Flowers: Solitary from axils of the leaves;
sepals 4, green, 2 small, 2 large; petals 4,
yellow, narrow; stamens yellow, many,
distinct. June through August.

Fruits: Capsule 2-to-4-valved, tawny-
brown; seeds small, black, numerous.

Distribution: Dry open woods.
Uncommon.

In Kentucky: AP, IP, ME.

This small shrub with cross-shaped
flowers is also called reclining St.
Andrew's cross and was named for St.
Andrew, who became the patron saint of
Scotland in the middle of the eighth
century. The generic name, *Hypericum,* is
from the Greek *hupereikon,* meaning
"above a picture," and refers to plants
that were hung over religious images or
pictures to ward off evil spirits when
St. John's-wort was in flower.

Native Americans used this plant
to cure horses that were bitten by
rattlesnakes.

Yellow flag iris
Iris Family
Iris pseudacorus L.
Iridaceae

Key features: Leaves swordlike; flowers showy, yellow with 3 large outer sepals and 3 narrow, upright petals.

Origin: Europe.

Life form: Perennial herb from stout rhizome.

Stems: Erect, smooth, shorter than, or equaling the leaves; to 3 feet tall.

Leaves: Blades, narrow, pale green, erect below, arching or nodding toward the tip; 1 to 3 feet tall.

Flowers: Subtended by a large leaflike bract; sepals 3, recurved, crest raised with brown lines or flecks; petals 3, erect, shorter than the sepals; stamens 3; style divided into 3 petal-like branches that arch over the stamens; 3 inches wide. May through June.

Fruits: Capsule oblong, green, slightly 3-angled; seeds corky, brown.

Distribution: Shallow water, mudflats. Rare.

In Kentucky: Collected in 10 counties. Listed as a Moderate Threat by the Kentucky Exotic Pest Plant Council.

Yellow flag iris was introduced into Canada from Eurasia and Africa and escaped from cultivation, spreading into freshwater wetlands. It has been cultivated in this country since Colonial times. The first record of this escapee dates from 1911 in Newfoundland. By 1915, the species was found in Nova Scotia and eventually spread into British Columbia, Quebec, Ontario, and Prince Edward Island. Currently, it has become an aggressive plant throughout most of the United States, growing in wetlands and in shallow water.

This iris is believed to be the source of the fleur-de-lis of the French heraldic shield, which dates from the twelfth century. The name *flag* is derived from *flagge,* which means "rush" or "reed."

Northern horse-balm

Mint Family
Collinsonia canadensis L.
Lamiaceae

Key features: Plant mildly lemon-scented; stems 4-angled; leaves opposite, toothed; flowers pale yellow, 2-lipped with 2 long stamens protruding and a long 2-forked style.

Origin: Native.

Life form: Perennial herb from a dark, knotty rhizome.

Stems: Erect, smooth, branched above; 2 to 3 feet tall.

Leaves: Upper sessile, blades broadly ovate, bases slightly heart-shaped to tapering, tips tapering; lower ones long-stalked, becoming progressively smaller; all distinctly veined; to 8 inches long.

Flowers: In loose terminal and axillary clusters; sepals 5, 2-lobed, ribbed; petals somewhat 2-lipped, upper 4 lobes nearly equal, lower lobe fringed, throat white-hairy inside; ½ to 1 inch long. July through September.

Fruits: Nutlets brown, rounded and smooth.

Distribution: Moist open woods. Uncommon.

In Kentucky: AP, IP, ME.

This member of the mint family is also called rich weed and toe itch. However, the more commonly used name, citronella, refers to the fact that the plant has a strong lemon scent. It was discovered and promoted by the famous English naturalist Peter Collinson (1693–1768), who took interest in cultivating new plants from America.

In parts of the Appalachian Mountains, this plant is called gravelroot or stoneroot because it was used to treat kidney stones.

Leaves: Evergreen, long-stalked except the uppermost; blades ovate, margins irregularly toothed, tips pointed to blunt; 1/2 to 2 1/2 inches long.

Flowers: In whorls; sepals 2-lipped; petals tubular, 2-lipped, upper hooded petal prominent, lower 3-lobed; stamens 4, ascending under the upper lip. May through June and sometimes in early fall.

Fruits: Nutlets.

Distribution: Moist open woods. There is an established colony at Iroquois in woods northeast of the stables and a small population in Cherokee in Wildflower Woods.

In Kentucky: (IP-rare) new state record. Listed on Watch List by the Kentucky Exotic Pest Plant Council.

Yellow lamium

Mint Family
Lamium galeobdolon (L.) Crantz
Lamiaceae
[*syn=Lamiastrum galeobdolon* (L.) L.]

Key features: Perennial ground cover with long runners; leaves opposite, dark green with whitish silver blotches above; flowers yellow, in axillary and terminal whorls.

Origin: Europe.

Life form: Perennial herb from extensive rhizomes.

Stems: Erect to trailing, tips of branches ascending, rooting at lower nodes, 4-angled, hairy; 8 to 24 inches tall.

This ornamental plant is often used as a groundcover in shady situations and has escaped from nearby urban gardens to become established in Cherokee and Iroquois. It spreads by stem fragments rooting at the node and seeds. Collections from Iroquois and Cherokee provided the first record of this species in Kentucky where it has been persisting in moist open woods. The dense silvery white mottled foliage is easily recognizable. Another common name is yellow archangel.

This species is related to the dead-nettles and said to hybridize with closely related taxa. It is invasive in the Pacific Northwest and also in several midwestern and eastern states.

Square-pod water-primrose
Evening-primrose Family
Ludwigia alternifolia L.
Onagraceae

Key features: Leaves alternate, narrow; flowers solitary, bright yellow, short-stalked in upper leaf axils; fruit capsules brown, squared above.

Origin: Native.

Life form: Perennial herb from fibrous roots.

Stems: Erect, diffusely branched, smooth or sparsely hairy, often reddish at flowering time; 2 to 3 feet tall.

Leaves: Sessile or short-stalked; blades linear-lanceolate, midvein distinct below, minute hairs on both surfaces, margins entire, tapering at both ends; to 3 inches long.

Flowers: Sepals 4, reddish green, ovate-triangular; petals 4, separate, shorter than sepals, falling early; stamens 4, alternating with petals; ¾ inch wide. June through September.

Fruits: Capsule, angles slightly winged, rounded at base; seeds smooth, plump.

Distribution: Wet, open ground at Summit Field. Common.

In Kentucky: AP, IP, ME.

Other common names are rattlebox and bushy seedbox, which aptly describes the seeds rattling in the squared fruit capsule. The genus is named for the great professor of botany in Leipzig, Christian Gottlieb Ludwig (1709–1773).

Creeping water-primrose

Evening-primrose Family
Ludwigia peploides (HBK.) P.H. Raven
subsp. glabrescens (Kuntze) Shinners
Onagraceae

Key features: Plant aquatic; stems creeping or floating, often forming dense mats; leaves alternate, short-stalked, shiny green; flowers bright yellow, solitary.

Origin: Native.

Life form: Perennial herb from stolons.

Stems: Rooting at the nodes in wet ground or in shallow water; to 24 inches long.

Leaves: Blades oblong-lanceolate to obovate, smooth, narrowed at the base into a stalk; leaves at tips usually larger than those below, often in a rosette; 2 to 4 inches long.

Flowers: On long stalks from leaf axils; sepals 5, green; petals 5, obovate, stamens 10; pistil 1, stigma 5-lobed. June through August.

Fruits: Capsule cylindrical, smooth; seeds in a vertical row.

Distribution: In shallow, still water, mudflats. Locally common in Cherokee Lake, some roadside ditches, and scattered along Beargrass Creek; rare at Iroquois.

In Kentucky: (AP-rare), (IP-scattered but especially common along Ohio River counties), (ME-common).

The genus name is in honor of Christian Gottlieb Ludwig, a professor of botany in Leipzig.

Common evening-primrose

Evening-primrose Family
Oenothera biennis L.
Onagraceae

Key features: Leaves alternate, simple, red-spotted; flowers bright yellow, stigma cross-shaped; in terminal clusters that lengthen with maturity.

Origin: Native.

Life form: Biennial herb from a taproot.

Stems: Stout, erect, single or branched; 1 to 5 feet tall.

Leaves: Basal rosette blades lanceolate, long-stalked; stem leaves alternate, blades lanceolate to oblong, margins wavy toothed, bases tapering, tips pointed; 4 to 8 inches long.

Flowers: Sepals 4, green, reflexed; petals 4, broad; stamens 8; pistil 1 with a cross-shaped stigma; 1 to 2 inches wide. June through September.

Fruits: Capsules cylindrical, 4-angled, velvety-hairy; seeds many, reddish brown, small, irregularly shaped.

Distribution: Disturbed ground, especially along roadsides, fields, thickets, limestone ledges.

In Kentucky: AP, IP, ME.

The first year's rosettes, as well as the roots, are used as greens and considered by some chefs as an essential ingredient in making tasty dishes. At one time, this species was cultivated for the roots, which are wholesome and nutritious. The Cherokee Indians boiled the roots like potatoes. They also made a root tea that was taken for obesity.

The Iroquois Indians used this herb to help cure laziness, and the roots were rubbed on the muscles of athletes to give strength.

The flowers open in the evening and close the following morning—hence the common name evening-primrose.

Flowers: Sepals 5, green with reddish dots, ovate-triangular; petals 5, yellow with reddish dots; stamens 5, unequal in length; pistil green, style about equaling stamens; 1 inch wide. June through September.

Fruits: Capsule round; seeds tiny, dark, 3-angled with scaly ridges.

Distribution: Disturbed ground in moist shady woods, waterways, low moist fields, turf, roadside ditches. Common.

In Kentucky: AP, IP, ME. Listed as a Severe Threat by the Kentucky Exotic Pest Plant Council.

Moneywort

Primrose Family
Lysimachia nummularia L.
Primulaceae

Key features: Plant creeping, rooting at the nodes; leaves opposite, rounded; flowers solitary in the leaf axils, petals bright yellow with reddish purple dots.

Origin: Europe.

Life form: Perennial herb from fibrous roots.

Stems: Creeping, often mat-forming, light green, smooth; to 3 feet long.

Leaves: Blades ovate-rounded, shiny with dark purple dots, margins entire, distinctly veined; 1 to 1½ inches long.

This ornamental ground cover from Europe has escaped cultivation and is now naturalized throughout the northeastern United States. It often forms new roots in the axils of the leaves and generates new stems that sprawl over the ground in all directions.

Many local names have originated with this global weed. The old genus name, *Serpentaria,* comes from a superstition that wounded snakes would crawl over and lie on this plant to heal. Other common names such as creeping Charlie, creeping Jenny, and wandering Taylor describe the habit, while *nummularia,* the species name, means "coin."

Hairy buttercup

Buttercup Family
Ranunculus sardous Crantz.
Ranunculaceae

Key features: Plant poisonous; basal leaves palmately 3-lobed, upper alternate, narrow; flowers solitary, bright yellow; fruits with knobby projections.

Origin: Europe.

Life form: Winter annual herb from a fibrous taproot.

Stems: Erect, several, freely branching, smooth to hairy; to 20 inches tall.

Leaves: Basal, long-stalked, blades irregular broadly triangular-ovate, margins entire, bases rounded, tips blunt; upper leaves alternate, smaller, narrow.

Flowers: Sepals 5, reflexed, green with white papery margins, soon falling; petals 5, obovate; stamens and pistils numerous; $\frac{1}{2}$ to 1 inch wide. May through July.

Fruits: Heads rounded, long-stalked; achenes many, round, brown with thin green border, knobby projections, beak short, curving upward.

Distribution: Moist ground in fields, roadside ditches, waterways. Locally common where escapes mowing.

In Kentucky: IP, ME.

This showy buttercup was collected in New Brunswick and Ontario, Canada, in the 1800s and has spread southward into most of the United States, where it is now a troublesome pasture weed.

The species name, *sardous,* means "bitter or scornful laughter." It is said that when this poisonous plant is eaten, the expression on the face is contorted, looking scornful.

Soft agrimonia

Rose Family
Agrimonia pubescens Wallr.
Rosaceae

Key features: Leaves alternate, pinnately divided, primary leaflets 3 to 9, with smaller leaf fragments growing between them; flowers yellow; fruit bristly.

Origin: Native.

Life form: Perennial herb from fibrous roots and fusiform tubers.

Stems: Light green, densely short-hairy, little branched above; 1 to 3 feet tall.

Leaves: Blades ovate to obovate, dull green above, whitish green below, hairy, margins coarsely toothed, bases wedge-shaped, tips pointed; leaflike stipules at base of compound leaf, toothed to deeply cleft, about ½ inch long.

Flowers: Small, remotely spaced on hairy, slender stems to 1 foot long; sepals 5, triangular; petals 5, yellow, oblong, spreading; stamens 5 to 20; about ¼ inch wide. July through August.

Fruits: Achenes 2 (3), enclosed by hard sepals with hooked bristles.

Distribution: Dry to moist open woods, woodland edges. Uncommon.

In Kentucky: AP, IP.

This species is one of four in the genus *Agrimonia* in Kentucky. The distinct fruits with hooked bristles are easily dispersed by animals and humans, readily attaching to fur or clothes. Even reports of bird dispersal, though rare, have been published.

Often growing in very dry areas, the plants have tuberous roots that are believed to store food and water.

The generic name is derived from the Greek *argemone,* meaning "poppy."

Common mullein
Figwort Family
Verbascum thapsus L.
Scrophulariaceae

Key features: Plant woolly; leaves basal and alternate; flowers bright yellow, in clusters on a dense, elongated spike with a few flowers opening at a time.

Origin: Europe.

Life form: Biennial herb from a taproot.

Stems: Stout, erect, unbranched or with a few upright branches near the top; 3 to 7 feet tall.

Leaves: Basal rosette, blades oblong, to 12 inches long, thick, distinctly veined, margins entire, tapered to a thick base; stem leaves alternate above, smaller, narrower, with leaf bases running down stem making it appear winged and 4-angled.

Flowers: In a cylindrical terminal spike to 2 feet long; sepals 5, green, hairy; petals 5; stamens 5, upper 3 hairy, lower 2 hairless; pistil 1; ¾ to 1 inch wide. June through September.

Fruits: Capsules round, hairy, style persistent; seeds tiny, ridged, deeply grooved.

Distribution: Disturbed dry ground in fields, roadsides, rocky ledges. Uncommon.

In Kentucky: AP, IP, ME.

Similar species: Moth mullein (*Verbascum blattaria* L.) from Eurasia, is a slender, biennial herb to 3 feet tall with button-like buds that open into **5 white to yellow rounded petals with hairy, purple stamens that bend downward.** The **leaves are small, smooth,** and to 4 inches long. The plant has glandular hairs throughout. Cher, Iroq, Shaw, Chick. Disturbed dry ground, fields, roadsides, thickets. Uncommon. June through July. In Kentucky: (AP-rare), (IP-mostly Bluegrass Region), ME.

Common mullein is also known as Jacob's staff and flannel-leaf. A folk term was candlewicks, so named because the hairs on the leaves were scraped off and used to make candlewicks. The name moth mullein comes from the resemblance of the curved fuzzy stamens to a moth's antennae.

In the fourth and fifth centuries BC, the flowers were used to make yellow hair dye. A soap made from the ashes of the leaves was said to restore graying hair to its original color.

Native Americans smoked the dried leaves and roots as a treatment for asthma and other pulmonary diseases.

Jerusalem-artichoke

Aster Family
Helianthus tuberosus L.
Asteraceae

Key features: Plant from long cord-like rhizomes with tubers at the tips; leaves opposite below, alternate above, leaf stalks winged; flower heads showy, yellow.

Origin: Native.

Life form: Perennial herb.

Stems: Stout, hairy, often branching above; 3 to 8 feet tall.

Leaves: Blades lanceolate to ovate, 3-nerved, rough above, margins toothed; 3 to 10 inches long.

Flowers: Terminal heads, 1-to-5-flowered; inner disk florets dark yellow, small, tubular; outer ray florets, yellow, 10 to 20; involucral bracts green, narrow, tips long tapering; 3 to 4 inches wide. August through October.

Fruits: Achenes golden to dark brown, flattened, 2-pronged; seed 1.

Distribution: Along waterways, open fields, woodland edges. Locally common along the River Walk at Shawnee; uncommon elsewhere.

In Kentucky: IP, AP, ME.

The tuberous rhizomes, which vaguely resemble ginger root or a small potato, are a valuable food source and contain high levels of potassium, phosphorus, and inulin. Folk uses have suggested that eating the edible rhizomes may aid in treating diabetes.

Native Americans, who called the plant sun roots, cultivated this plant even before the arrival of the early settlers and used the rhizomes for food. It wasn't until the sixteenth century that it reached Europe and was cultivated in many parts of the world, although it remains a minor crop. One plant is capable of producing 75 to 200 tubers in a single growing season, and there can be as many as six shoots from one tuber.

Peeled and sliced, the tubers are used to make a cold salad, similar to potato salad, or they can be steamed and eaten like a vegetable. The flavor is said to be similar to that of artichoke, although they are not related.

Axillary goldenrod

Aster Family
Solidago caesia L.
Asteraceae

Key features: Stems often arching to one side; leaves alternate, narrow; flower heads small, yellow, in axillary clusters at the base of the leaf.

Origin: Native.

Life form: Perennial herb from stout rhizomes.

Stems: Solitary to branched, bluish gray to purple with a whitish bloom, smooth; 1 to 3 feet tall.

Leaves: Alternate, blades lanceolate to elliptic, upper surface dark green, slightly hairy, margins finely toothed, midvein distinct, tapering to a sessile base, tips pointed; 2 to 5 inches long.

Flowers: Inner disk florets 5 to 7, yellow; outer ray florets, 3 to 5, yellow; involucral bracts green, narrow, rounded at tip, obscurely nerved. August through October.

Fruits: Achenes hairy.

Distribution: Open dry to moist woods, woodland edges, shady roadsides. Common.

In Kentucky: AP, IP, ME.

This graceful goldenrod has a distinct bluish gray stem covered with a whitish bloom and offsets the small yellow flower clusters in the upper leaf axils, giving it a slender wand-like appearance. It is also called wreath goldenrod.

Common goldenrod

Aster Family
Solidago canadensis L.
Asteraceae

Key features: Plant from rhizomes, often forming dense colonies; stems densely hairy; leaves alternate, narrow, rough; flowers yellow, on one side of backward-curving branches.

Origin: Native.

Life form: Perennial herb.

Stems: Stout, erect; 2 to 6 feet tall.

Leaves: Numerous, crowded, smaller upward; blades lanceolate to elliptic, sessile, 3-nerved, margins with small teeth, tapering at both ends; 3 to 6 inches long.

Flowers: In small heads; inner disk florets yellow, 2 to 8; outer ray florets yellow, mostly 8, slender, short; involucral bracts yellowish without a well-defined green tip, thin, slender. August through October.

Fruits: Achenes longitudinally ridged, short-hairy, tufts of hairs at apex.

Distribution: Roadsides, woodland edges, thickets, fields. Abundant.

In Kentucky: AP, IP, ME.

Similar species: Smooth goldenrod (*Solidago gigantea* Aiton) is a native perennial herb growing to 7 feet tall and difficult to distinguish from the above species. The **stems are greenish purple, smooth, with a whitish bloom that can be rubbed off** and **leaves that are smooth** with hairs on the veins below. The small flowers are on one side of backward-curving branches. There are 7 to 15 outer yellow ray florets and 5 to 11 inner yellow disk florets; involucral bracts blunt, in 2 to 5 overlapping series. This species flowers earlier than common goldenrod and prefers moist to wet soils. Cher, Sen, Iroq, Shaw, Chick. Common. July through October. In Kentucky: AP, IP, ME.

Common goldenrod is the most abundant goldenrod species in Kentucky, although the taxonomy is confusing because of the variations within the species.

The foliage of common goldenrod contains a volatile oil with a fragrance similar to that of pine needles. The roots have chemicals that inhibit the growth of other plants. Even a single plant can form a dense thicket from a web of roots that spreads out 360 degrees from the parent plant.

The roots were used by Native Americans to treat burns. A decoction was used as a bath for babies with excessive crying and sleeplessness. It was also mixed with wild tarragon and used as a wash for horses with cuts and sores.

Elm-leaved goldenrod

Aster Family
Solidago ulmifolia Muhl.
Asteraceae

Key features: Stems arching, widely spreading, slightly drooping; leaves alternate, smooth; flower heads yellow, crowded on one side of branch.

Origin: Native.

Life form: Perennial herb from fibrous roots and rhizomes.

Stems: Slender, mostly single, green and tan ridged, hairs mostly curled; to 4 feet tall.

Leaves: Lower leaves abruptly tapering into a stalk, soon deciduous; blades lanceolate to ovate, distinctly veined, margins entire to toothed, bases tapering into a short stalk, tips pointed; reduced upward, sessile; 1/4 to 4 inches long.

Flowers: In a few elongated branches; inner disk florets 4 to 7, tubular; outer ray florets 3 to 5, minute; involucral bracts in 2 series, green-tipped. August through October.

Fruits: Achenes small, short-hairy.

Distribution: Moist to dry woods, woodland edges, fields, roadsides. Common at Iroquois; rare elsewhere.

In Kentucky: (AP-rare), (IP-mostly Outer Bluegrass), ME.

The common name refers to the outward spreading flowering branches that resemble the shape of an elm tree.

Wingstem

Aster Family
Verbesina alternifolia (L.) Britton
Asteraceae

Key features: Leaves alternate, narrow, leaf bases tapering into wings down the stem; flower heads with greenish yellow center surrounded by drooping yellow ray florets of unequal length.

Origin: Native.

Life form: Perennial herb from rhizomes.

Stems: Erect, leafy, multiple from base, strongly winged, white hairs between the ridges; 3 to 8 feet tall.

Leaves: Blades lanceolate to lance-elliptic, rough, margins smooth to slightly toothed, white hairs on veins below, tips pointed; 4 to 10 inches long.

Flowers: Heads many, in open terminal clusters; inner disk florets greenish yellow, projecting outward forming a circle; outer ray florets 2 to 10; stamens 5, anthers purple; involucral bracts few, green, narrow, the outer soon reflexed; 1 to 3 inches wide. August through October.

Fruits: Achenes flat, ovate, spreading in all directions forming a rounded head.

Distribution: Low open woods, waterways, roadsides, thickets. Common.

In Kentucky: AP, IP, ME.

Similar species: Frostweed (*Verbesina virginica* L.) is a stout, native perennial herb to 8 feet tall. The alternate, greenish yellow leaves have winged leaf bases and the flowers are borne in flat-topped clusters with many **inner white disk florets and a few outer white ray florets.** Cher, Sen, Iroq. Along waterways, fields. Uncommon. August through October. In Kentucky: IP, ME.

Wingstem is also called yellow ironweed because of its resemblance to those species in the genus *Vernonia,* or ironweeds. Native Americans used the root bark as a purification emetic after the death of a patient. An infusion was also used to treat eye disease, fever, headache, and chills.

Wild geranium

Geranium Family
Geranium maculatum L.
Geraniaceae

Key features: Leaves palmately 5-to-7-lobed with coarsely toothed segments; flowers pinkish purple with purple veins; fruits slender, long, like a "cranes bill."

Origin: Native.

Life form: Biennial herb from stout rhizomes.

Stems: Erect, slender, hairy; 1 to 2 feet tall.

Leaves: Basal, few, long-stalked, palmately lobed, upper leaf surface often mottled in green and purple, bases wedge-shaped; stem leaves similar, short-stalked, smaller, 3 to 5 inches wide.

Flowers: One to 5, in loose axillary clusters; sepals 5, green, long-tipped; petals 5; stamens 10; 1 to 2 inches wide. April through May.

Fruits: Capsule splits into 5 slender, beak-like parts that coil upward and remain attached at the tip; seeds tiny.

Distribution: Moist open woods. Rare.

In Kentucky: AP, IP, ME.

Similar species: Carolina cranes-bill (*Geranium carolinianum* L.) is a native winter annual that grows to 2 feet tall. The **leaves are deeply palmately cleft or lobed, each with 5 to 9 narrow segments.** The small, 5-petalled **white to pink flowers,** are borne in clusters at the tips of the stem. The fruits are also long and slender, resembling a "cranes bill." Cher, Sen, Iroq, Shaw, Chick. In fields, roadsides, woodland edges, limestone ledges, cultivated beds. Common weed. April through July. In Kentucky: AP, IP, (ME-rare).

The generic name comes from the Greek word *geranos* "crane" and refers to the resemblance of the slender fruit to long-billed birds.

The rhizomes of wild geranium contain a high concentration of tannins and, when dried, produce a brownish powder known for its astringent properties. In Appalachia, a tea was used to treat dysentery and sore throat.

Although cultivated geraniums belong to the genus *Pelargonium,* they are in the same family as the wild species.

Red dead-nettle

Mint Family
Lamium purpureum L.
Lamiaceae

Key features: Stems square; leaves opposite, heart-shaped, bluntly toothed; flowers pinkish purple in whorls in upper leaf axils.

Origin: Eurasia.

Life form: Winter annual with fibrous roots.

Stems: Erect to sprawling and rooting at the nodes, branched, purplish; to 16 inches tall.

Leaves: Lower ones long-stalked, upper similar, short-stalked; blades broadly ovate, purplish, overlapping, angled downward; ½ to 1 inch long.

Flowers: Sepals tubular, teeth 5, long-pointed; petals 2-lipped, upper lip arched, lower lip purple-spotted, tip notched; ½ to ¾ inch long. February to May and sporadically in fall.

Fruits: Nutlets 4, tiny, brown, mottled white.

Distribution: Disturbed ground in turf, moist woods, roadsides, fields, waterways, cultivated beds. Common.

In Kentucky: AP, IP, ME. Listed as a Moderate Threat by the Kentucky Exotic Pest Plant Council.

Similar species: Henbit (*Lamium amplexicaule* L.) is a winter annual weed from Eurasia with **opposite, clasping leaves that are circular in outline and form a "collar" around the square stem;** lower ones are similar but long-stalked. The small **pinkish purple flowers** are 2-lipped and produced in **whorls in the upper leaves.** Cher, Sen, Iroq, Shaw, Chick. Found in similar habitats as the above species and often growing together. February to May and sporadically in fall. In Kentucky: AP, IP, ME. Listed as a Moderate Threat by the Kentucky Exotic Pest Plant Council.

The generic name is derived from the Greek *lamos* "throat" and refers to the wide-open throat of the corolla.

Henbit is found throughout the world and has survived all the major changes in

Invasive plant
Cher, Sen, Iroq, Shaw, Chick

agriculture for more than 150 years. It is highly adaptable and opportunist and has the ability to form adventitious roots at the lower nodes, produce seeds readily, and develop slowly throughout the winter months in many areas.

Both species are common weeds growing throughout Kentucky.

Virginia spring-beauty

Purslane Family
Claytonia virginica L.
Portulacaceae

Key features: Plant from underground corms; leaves opposite, stalked, usually a single pair produced midway up the stem; flowers pink or white with darker pink veins, produced in a floppy raceme.

Origin: Native.

Life form: Perennial herb.

Stems: Weak, succulent, light green to reddish; 4 to 10 inches tall.

Leaves: Blades narrow, fleshy, central midvein distinct running the length of leaf, tip long-tapering; 2 to 6 inches long.

Flowers: Several, about ½ inch wide; sepals 2, green; petals 5; stamens 5, anthers pink; style 3-parted. March through May.

Fruits: Capsule ovoid, enclosed by 2 persistent sepals.

Distribution: Moist woods, shady turf. Common at Cherokee; locally common at Shawnee, Chickasaw, and Seneca along roadsides and in turf under shade trees; uncommon in Iroquois.

In Kentucky: AP, IP, ME.

The genus *Claytonia* was named in honor of Dr. John Clayton, a distinguished physician who came to Virginia in 1705 to collect medicinal plants. Being at the forefront of medical botany in the eighteenth century, his specimens are known in both herbaria and gardens in Europe and America. He also contributed to the *Flora of Virginia*, published in 1739 and 1743.

The underground corms when boiled in salted water are nutritious and taste like chestnuts. Native America children likened these tasty corms to nuts and considered them "candy."

Red columbine
Buttercup Family
Aquilegia canadensis L.
Ranunculaceae

Key features: Leaves divided into 3 leaflets that are mostly 3-lobed, the margins irregularly cut; flowers yellow with a red spur extending backward.

Origin: Native.

Life form: Perennial herb from a thick rhizome.

Stems: Slender, branched, reddish green; to 3 feet tall.

Leaves: Basal and stem leaves similar, long-stalked below, sessile upward; blades greenish white above, bases wedge-shaped; up to 3 inches long.

Flowers: Single or in groups of 2 or 3, nodding on long stalks; sepals 5, red, ovate; petals 5, funnel-shaped, yellow above, nectar-tipped spurs red, slender, hollow; stamens yellow, many, forming a central column with the 5 pistils. April through May.

Fruits: Pod tubular, splitting open along 1 side; seeds many, shiny and black.

Distribution: Dry to moist woods, limestone ledges. Uncommon at Cherokee: rare in Seneca.

In Kentucky: AP, IP.

The genus name *Aquilegia* comes from either *aquila* "eagle" or *aqu* "water" and *lego* "to gather"; this refers to the drops of nectar that can be seen in the falcon-like spurs. The five spurs are also said to resemble a circle of doves drinking around a fountain.

Another name is rock bells and describes the plant's common habitat.

Native Americans made a pulverized paste from the seeds that they placed among blankets or clothing and used as perfume. The Iroquois Indians made an infusion with other plants that was used as a wash for poison-ivy.

Toadshade trillium

Trillium Family
Trillium sessile L.
Trilliaceae

Key features: Leaves in a single whorl of 3, green or mottled in various shades of green; flowers solitary, sessile, with 3 maroon, narrow petals.

Origin: Native.

Life form: Perennial herb from thick rhizomes.

Stems: Lacking except for the flowering stalk to 8 inches tall.

Leaves: Leaflike bracts 3, whorled, sessile, blades ovate to elliptic, margins entire, tips short pointed; 2 to 4 inches long.

Flowers: Sepals 3, lanceolate, spreading or ascending; petals 3, erect; stamens 6, half as long as the petals; ovary 6-angled, greenish white below, purple above; about 1 inch long. March through May.

Fruits: Berrylike, dark greenish purple, 6-angled, slightly winged.

Distribution: Moist to dry woods. Common in Cherokee; uncommon in Seneca; rare in Iroquois.

In Kentucky: (AP-rare), IP.

This beautiful wildflower is the symbol for modest beauty. It was used traditionally by women as a love potion where the boiled roots were dropped into the food of the desired. Native Americans soaked the roots and made an eye wash.

It is also called sessile trillium because of the direct attachment of the leaves and flowers to the stem. This species emits an odor comparable to that of a dead animal and attracts flies and beetles as pollinators. Because of the raw beef scent, it is also known as bloody butchers.

Field garlic
Onion Family
Allium vineale L.
Alliaceae

Key features: Plant bulbous; odor of garlic or onion; leaves grasslike; flowers pinkish purple, in umbels, or sometimes replaced by small bulbils.

Origin: Europe.

Life form: Perennial herb from aerial and underground bulblets.

Stems: Slender, solid, unbranched, the lower halves covered with membranous leaf bases that become stiff with age; 1 to 3 feet tall.

Leaves: Basal and alternate; floppy above, hollow, smooth, bluish green, cylindrical, tapering to a tip.

Flowers: In rounded, terminal umbels subtended by a papery sheath-like spathe; flowers, when present, with 6 tepals; stamens 6; pistil 1; aerial bulbils small, produced in place of flowers and develop green, tail-like leaves about 1 inch long. May through June.

Fruits: Capsules oval, 3-parted; seeds 2, dull black, tiny, convex on 1 side, with bumps.

Distribution: Disturbed ground, in fields, turf, cultivated beds, waterways. Common.

In Kentucky: AP, IP, ME.

The generic name *Allium* includes such important cultivated members as onion (which exists only in cultivation), garlic, shallot, chive, scallion, and leek.

Field garlic has become one of the most noxious weeds to be introduced into the middle and southeastern United States. Early records show that it was a serious problem in this country more than a century ago. Milk from livestock that have grazed on it has a garlic-like flavor, which decreases its salability.

Early physicians wore garlic bulbs on strings around their necks to project them from getting the diseases of their patients.

Swamp milkweed

Milkweed Family
Asclepias incarnata L.
Asclepidaceae

Key features: Plant with milky sap; leaves opposite, narrow, entire; flowers small, pink to purplish pink, in terminal and axillary umbels; fruit pods in pairs.

Origin: Native.

Life form: Perennial herb from a taproot and rhizomes.

Stems: Stout, erect, unbranched to much branched; to 5 feet tall.

Leaves: Blades linear-lanceolate to elliptic, midvein distinct below, bases rounded, squared, or slightly heart-shaped, tips tapering or short-pointed; 2 to 6 inches long.

Flowers: Sepals 5, lanceolate to ovate, green to purple; petals 5, bent downward; center with 5 white hoods and a thin recurved horn extending above. June through September.

Fruits: Pods 2 to 4 inches long, erect, smooth, opening on 1 side, on straight or curved stalks; seeds with tufts of silky white hairs.

Distribution: Wet to moist ground. Common.

In Kentucky: (AP-rare), IP, ME.

The genus was named by Carl Linnaeus, the great Swedish botanist, in honor of Homer's physician, Asclepios, who became the guardian god of medicine.

The Iroquois Indians dried the stems to make cord used for extracting teeth. The fibers were also made into strong fishing lines and sewing threads. The silky hair from the fruit pods was used to make beds and pillows considered far better than those made with feathers and cotton.

While the floral scent is said to resemble that of cinnamon, the foliage is toxic and contains cardiac glycosides.

Bull thistle

Aster Family
Cirsium vulgare (Savi) Ten.
Asteraceae

Key features: Stems stout with irregular spiny wings; leaves pinnately divided, spine-tipped; flower heads pinkish violet, solitary, constricted between the bracts and flowers.

Origin: Eurasia.

Life form: Biennial herb from a taproot.

Stems: Woody, ridged, branched, hairs cobwebby; 2 to 6 feet tall.

Leaves: Basal rosette; blades oblong-lanceolate, 3 to 12 inches long, deeply divided into narrow lobes ending in a pale yellow spine, margins and upper leaf surface spiny, white-woolly below; stem leaves alternate, similar, reduced upward, leaf bases clasping, extending downward forming a winged stem.

Flowers: Many, in dense terminal heads, 1½ to 2½ inches wide; inner disk florets small, tubular; involucral bracts green, reflexed, hairs cobwebby, spines yellow, needlelike. June through September.

Fruits: Achenes oval, glossy yellowish brown with darker stripes and tufts of soft white branched hairs at tip; 1-seeded.

Distribution: Disturbed ground in fields, roadsides, thickets, open woods, waterways. Common.

In Kentucky: AP, IP, ME. Listed as a Significant Threat by the Kentucky Exotic Pest Plant Council.

Similar species: Musk thistle (*Carduus nutans* L.) is an invasive biennial herb from Eurasia. The stout, spiny, branched stems to 4 feet tall have deeply pinnately lobed leaves with spine-tipped margins and leaf bases forming a winged stem. The **solitary, showy, pinkish purple flower head nods at the tip.** Below there are many greenish to purplish spine-tipped bracts and **the flower head is not constricted between the bracts and flowers**. Cher, Sen, Iroq, Shaw, Chick. Disturbed ground. Uncommon. June through September. In Kentucky: AP, IP, ME. Listed as a Severe Threat by the Kentucky Exotic Pest Plant Council.

Bull thistle is also called spear thistle, which describes the sharp-pointed spines on the stems and leaves. It is especially troublesome in western North America, where it is a problem on rangelands and in eroded gullies, ditches, fencerows, and fertile agricultural lands. The seeds are carried by wind and water, but they are transported in large quantities from the movement of baled or loose hay or forage.

Musk thistle was brought to the attention of the University of Kentucky College of Agriculture in 1941 after it was sited growing along the Warren–Logan county line. It was said to have been brought in with contaminated bales of hay from the Midwest. Even earlier, this plant was first reported growing in a pasture in Indiana in 1934 and had been in that same locality for seventeen years and spreading since there was no effort to eradicate it at the time. It is considered to be one of the worst weeds in Kentucky.

Purple coneflower

Aster Family
Echinacea purpurea (L.) Moench.
Asteraceae

Key features: Basal leaves ovate, long-stalked, upper lanceolate, stalkless; flower heads terminal, solitary or few, ray florets pinkish purple, drooping.

Origin: Native.

Life form: Perennial herb from a thick rhizome.

Stems: Simple to slightly branched above, smooth; to 4 feet tall.

Leaves: Alternate above, blades lanceo-late, surfaces rough, green, veins 3 to 5, margins sharply toothed, bases rounded, tips pointed; 4 to 7 inches long.

Flowers: Inner disk florets many, rust-colored, domed; ray florets 15 to 20; involucral bracts narrow, overlapping, margins short-hairy; 2½ to 5 inches wide. June through October.

Fruits: Achenes thick, 4-sided, with small teeth.

Distribuion: Woodland edges, meadows in Cherokee; Summit Field at Iroquois. Uncommon.

In Kentucky: AP, IP.

Native Americans would chew the root as a remedy for coughs and dyspepsia. A simple or compound infusion of the root was effective for treating gonorrhea. It is also said to boost the body's immune system, and purple coneflower pills are a popular item sold in health food stores today.

Cultivated for commercial use, this beautiful wildflower is a popular plant used in native landscape gardens.

Purple-node joe-pye weed

Aster Family
Eupatorium purpureum L.
Asteraceae

Key features: Stems solid, green- or purple-tinged, nodes dark purple; leaves in whorls of 3 to 5; flower heads with 3 to 7 purplish pink flowers produced in dome-shaped terminal clusters.

Origin: Native.

Life form: Perennial herb.

Stems: Stout, usually not branched; to 8 feet tall; freshly bruised plants are strongly vanilla-scented.

Leaves: Blades lanceolate to ovate, dull green above, veined, margins sharply toothed, base tapering into a long or short stalk, tips pointed; 3 to 6 inches long.

Flowers: Disk florets small, tubular, style white, divided; involucral bracts in several series, pinkish purple, 3-nerved. July through September.

Fruits: Achenes black.

Distribution: Moist ground along waterways, ponds, fields. Uncommon. A showy, purple-stemmed cultivar "Gateway" has been planted at Breckenridge Springs.

In Kentucky: AP, IP, ME.

Similar species: Hollow-stemmed joe-pye weed (*Eupatorium fistulosum* Barratt) is a native plant to 8 feet tall. It has a **hollow, purplish stem with a whitish bloom** and leaves in **whorls of 4 to 7** that are lanceolate. The **flower heads are purplish pink with 5 to 8 flowers that form a round-topped or high-dome cluster.** There is no vanilla scent when the plant is bruised. Cher. Uncommon. July through September. In Kentucky: AP, IP, ME.

Purple-node joe-pye weed is also called Indian gravelroot and kidneyroot and was used in traditional medicine to treat urinary disorders. The Native Americans also use the fruit for a red or pink dye.

Another name for this group of plants is Queen-of-the-meadow.

Sessile blazing-star
Aster Family
Liatris spicata (L.) Willd.
Asteraceae

Key features: Plant from a corm; leaves narrow, crowded; flowers pinkish purple, 6 to 10 in heads crowded on an elongated terminal spike; bracts with blunt tips.

Origin: Native.

Life form: Perennial herb.

Stems: Stiff, erect, single or few from base, ridged; 2 to 5 feet tall.

Leaves: Alternate, sessile, linear, lower ones widest, veins parallel, margins entire, smaller above; 4 to 8 inches long.

Flowers: Small, disk florets tubular, 5-lobed; style whitish pink, divided, curled, extending beyond tube; involucral bracts green to purple tinged, sticky, flat. July through September.

Fruits: Achenes light brown, bristly.

Distribution: Fields and thickets at Cherokee; Summit Field at Iroquois. Uncommon.

In Kentucky: (AP-western part), IP, ME.

Similar species: Plains blazing-star [*Liatris squarrosa* (L.) Michx.] is a native plant to 3 feet all. The **few flower heads are pink** and the **involucral bracts** below have **long tips that are spreading to recurved.** Iroq. Dry rocky upland woods. Rare. July through September. In Kentucky: AP, IP, ME.

Several members of this genus are cultivated for their unique fuzzy purple flowers and distinct, often-colored bracts below.

Deer potato and button snake-root are other common names and refer to the rounded underground corms that were dug and stored in the fall and used in winter for making nutritious food by the Native Americans. The chewed corm was blown into the nostrils of horses to give them strength to run long distances without getting out of breath.

Cardinal-flower

Bellflower Family
Lobelia cardinalis L.
Campanulaceae

Key features: Stems erect; leaves alternate, narrow; flowers deep red, 2-lipped, clustered on a terminal raceme to 20 inches long.

Origin: Native.

Life form: Perennial herb from rhizomes.

Stems: Single, unusually unbranched, short-hairy, angled, purplish green; 2 to 3 feet tall.

Leaves: Short-stalked below, upper sessile; blades lanceolate to elliptic, short-hairy, margins irregularly toothed, base tapering, tips pointed; to 6 inches long.

Flowers: Sepals green, 5, narrow, flaring; petals 5, 2-lipped: lower lip 3-lobed, upper lip 2-lobed; stamens 5, filaments red, united into a tube surrounding a pistil that projects upward through slit in upper lip; 1 to 2 inches long. July through September.

Fruits: Capsule broadly bell-shaped; seeds brown, rough, narrow.

Distribution: Moist to wet ground along waterways, roadside ditches, swamps. Rare.

In Kentucky: AP, IP, (ME-rare).

This strikingly beautiful native wild-flower can be seen in late summer growing along creeks, pond margins, or meadows. Available in the nursery trade, the deep red flowers make an attractive display when planted in formal gardens or in natural areas. The nectar is a favorite of ruby-throated hummingbirds and swallowtail butterflies.

Some Native Americans used the plant for a love medicine in which the roots and flowers acted as a love charm. The plant is considered toxic.

The common name stems from the resemblance to the scarlet robes of the cardinals of the Roman Catholic Church. The generic name honors Matthias de L'Obel, a Flemish herbalist.

Crown-vetch

Legume Family
Coronilla varia L.
Fabaceae

Key features: Stems often forming a mound; leaves alternate, pinnately compound; flowers pink to pinkish white, pealike, in rounded axillary clusters.

Origin: Mediterranean.

Life form: Perennial herb from rhizomes.

Stems: Ascending or straggling on the ground to 6 feet, or forming a mound, 1 to 2 feet tall; tendrils absent.

Leaves: Sessile to short-stalked; odd-pinnate, leaflets 11 to 25, opposite, blades oblong to obovate, margins entire, tips blunt to shallowly notched or with a short, thin tip; 2 to 4 inches long.

Flowers: Clusters on slender stalks to 5 inches long; sepals tiny, bell-shaped, 5-lobed; petals 5, free, upper petal rounded, lower incurved, side petals 2; stamens 10. May through September.

Fruits: Pods narrow, 4-angled with 3 to 7, 1-seeded segments.

Distribution: Disturbed ground, along roadsides, fields, thickets, cultivated beds. Common.

In Kentucky: AP, IP ME. Listed as a Severe Threat by the Kentucky Exotic Pest Plant Council.

This ornamental plant was introduced into the United States from Europe. In the 1950s, it was used as a quickly growing ground cover for erosion control along roadsides, on steep banks, and as a cover crop in agriculture and mine reclamation. It is a prolific seed producer and spreads by both seeds and rhizomes.

The plant contains a toxic cardiac alkaloid, coronillin, and is potentially fatal if eaten.

The genus name comes from *corona*, "a crown," and refers to the arrangement of the flower clusters.

Panicled tick-trefoil

Legume Family
Desmodium paniculatum (L.) DC.
Fabaceae
[*syn=Desmodium glabellum* (Michx.)
DC.; *Desmodium perplexum* B.G. Schub]

Key features: Stems usually multiple from base; leaves alternate, 3-foliate, 3 times as long as wide; flowers pealike, pink, in loosely branched clusters.

Origin: Native.

Life form: Perennial herb from taproot.

Stems: Single or multiple, slender, erect, ascending or reclining, smooth or with a few hooked hairs; 2 to 3 feet tall.

Leaves: Blades lanceolate to ovate, terminal leaflet largest, lateral 2 smaller, margins entire, bases slightly unequal; 2 to 4 inches long; leaflike stipule at base of leaf stalk small, very narrow, long-tapering.

Flowers: Sepals 5, reddish; petals 5, pink, upper petal notched at tip with two yellowish spots near base; $\frac{1}{2}$ inch wide. July through September.

Fruits: Segments 3 to 4, triangular to rounded, veins netlike; seeds 1 per segment.

Distribution: Roadsides, thickets, woodland edges, fields. Common.

In Kentucky: AP, IP, ME.

Similar species: Naked tick-trefoil [*Desmodium nudiflorum* (L.) DC] has **two kinds of stems: a shorter sterile branch,** 3 to 12 inches tall, bearing a cluster of 3-foliate leaves at the summit and a **long, fertile, leafless stem** to 40 inches tall that bears the **pink pealike** flowers. Iroq. Open woods, woodland edges. Uncommon. July through September. In Kentucky: AP, IP, ME.

Trying to correctly identify the many species of tick-trefoils found throughout Kentucky can be difficult. They differ mainly in the shape of the leaves and fruit, which is called a loment. The latter is especially helpful in identification.

The generic name is derived from the Greek *desmos* "chain," which describes the jointed triangular seedpods that are covered with minute hooks and readily stick to animal fur and human clothing.

Smooth trailing lespedeza

Legume Family
Lespedeza repens (L.) Barton
Fabaceae

Key features: Plant trailing; stems smooth or with flattened hairs; leaflets 3, terminal one largest; flowers small, pinkish purple, pealike.

Origin: Native.

Life form: Perennial herb from a taproot and rhizomes.

Stems: Usually several from the base, often mat-forming; to 20 inches long.

Leaves: Alternate, compound; blades ovate to oblong, dull green above, grayish green below, hairs appressed, midvein distinct, tip short, needlelike; ½ to 1½ inches long.

Flowers: Few, in axillary clusters on a slender stalk; sepals 5, lobes sharp-pointed, hairy; petals 5, uppermost petal largest; to ¼ inch long; June through September.

Fruits: Small, broadly ovate, flat, purple-veined, short-beaked.

Distribution: Dry woods, roadsides, fields. Common.

In Kentucky: AP, IP, ME.

Similar species: Downy trailing lespedeza (*Lespedeza procumbens* Michx.) is a perennial herb from a taproot and rhizomes. The **trailing stems are densely white-hairy and often form mats on the ground.** The 3-foliate leaflets are oval to oblong and the pinkish purple pealike flowers are produced in few-flowered axillary clusters. Iroq. Dry woods, roadsides, fields. Common. August through October. In Kentucky: AP, IP, ME.

Both species are very similar except one has smooth stems and the other hairy; the common names describe this character. Trailing along the ground, the flowers contain nectar and are visited by bees.

Common marsh-pink
Gentian Family
Sabatia angularis (L.) Pursh.
Gentianaceae

Key features: Stems strongly 4-angled; leaves opposite above, clasping; flowers rose-pink, sometimes white, with a yellowish green center outlined in darker pink.

Origin: Native.

Life form: Biennial herb from fibrous roots.

Stems: Erect, branching above, smooth, winged; 1 to 2½ feet tall.

Leaves: Basal rosette below, blades obovate; opposite above, clasping, blades ovate to lanceolate, veins 3 to 7, margins smooth, tips pointed; 1 inch long.

Flowers: In terminal clusters; sepals green, linear-lanceolate; petals 5, obovate; stamens 5, anthers yellow; 1 to 1½ inches long. July through September.

Fruits: Capsule rounded; seeds tiny, pitted.

Distribution: Summit Field. Rare.

In Kentucky: AP, IP, ME.

The Gentian family contains many species with medicinal "bitter principles" found in the rhizome. This exquisite wildflower is also called bitter-broom because it was used by the early settlers as a bitter tonic. The Cherokee Indians made a tea that helped relieve menstrual pain. In France, the still popular brand of bitters "Suze," invented in 1885, is made by distilling gentian roots.

The generic name, *Sabatia,* honors the eighteenth-century Italian botanist Liberato Sabbati.

Wild bergamot

Mint Family
Monarda fistulosa L.
Lamiaceae

Key features: Plant aromatic; stems 4-angled; leaves opposite, ovate; flowers many, pink to lavender, in dense terminal rounded heads.

Origin: Native.

Life form: Perennial herb from fibrous roots and rhizomes.

Stems: Stout, light green to purplish green, branching above; 1 to 5 feet tall.

Leaves: Blades rough, margins toothed, bases slightly heart-shaped to squared, tips pointed; to 5 inches long.

Flowers: Heads 2 inches wide; petals tubular, 2-lipped, upper lip erect, toothed, folded around 2 protruding stamens and style; lower lip 3-lobed; involucral bracts, pale green, often pink-tinged. June through September.

Fruits: Nutlets 4, small, dark.

Distribution: Woodland edges, thickets, open fields; especially common at Summit Field in Iroquois; uncommon at Cherokee.

In Kentucky: AP, IP, ME.

The foliage of this attractive aromatic native plant is a source of the oil of thyme and the antiseptic "thymol," a primary ingredient in commercial mouthwashes.

Some Native Americans were known to recognize four varieties with different scents. An infusion made from the leaves or root was drunk and wiped on the head as a remedy for nosebleeds, headaches, colds, and bronchial problems. The root was used in a decoction and drank to help reduce pain in the stomach and the intestines.

Canada germander

Mint Family
Teucrium canadense L.
Lamiaceae

Key features: Stems 4-angled; leaves opposite, narrow; flowers pinkish white with darker purple lines near base of lower lip and 4 stamens protruding.

Origin: Native.

Life form: Perennial herb from fibrous roots and rhizomes.

Stems: Erect, stout, white-glandular-hairy, axillary branches smaller; to 3 feet tall.

Leaves: Lower short-stalked, upper sessile; blades lanceolate to ovate, margins toothed, distinctly veined above, white-hairy below, bases rounded to tapering, tips pointed; 2 to 5 inches long.

Flowers: Two to 6, in whorls arranged in a dense terminal spike; sepals usually 5, velvety hairy; petals 2-lipped with 5 lobes; stamens 4, exserted through a cleft between the 2 upper lobes, arched downward with style; ½ inch wide. June through September.

Fruits: Nutlets 4, round, reddish brown, wrinkled.

Distribution: Moist to wet open woods, woodlands edges, waterways. Uncommon.

In Kentucky: AP, IP, ME.

Similar species: Smooth hedge-nettle (*Stachys tenuifolia* Willd.) is a perennial plant with a square stem to 3 feet tall. The leaves are opposite, **oblanceolate**, smooth, with toothed margins and a pointed tip. The **pink to white 2-lipped flowers** are in clusters of 6 in the axils of the upper leaves and have **4 stamens included under the upper hooded lip.** Cher. Moist ground along roadsides, waterways, woodland edges. Uncommon. June through August. In Kentucky: AP, IP ME.

Canada germander is a summer wildflower that is found in the Bluegrass Region and is sold in the horticultural trade because of its showy flowers. It is also known as wild basil and wood sage and was used traditionally for lung ailments, worms, and piles.

The genus name, *Teucrium,* means "of Teucer," referring to the first King of Troy. *Stachys* is Greek for "spike" and describes the arrangement of the flowers within this large group of plants, also called betony or woundwort. Both plants are available in the nursery trade and often used in naturalistic landscapes.

Purple loosestrife
Loosestrife Family
Lythrum salicaria L.
Lythraceae

Key features: Leaves opposite or whorled, narrow; flowers pinkish purple, in clusters on dense terminal spikes to 1 foot long.

Origin: Eurasia.

Stems: Erect, woody, multiple from base, 4-angled, rough, hairy; 3 to 5 feet tall.

Leaves: Blades lanceolate-oblong, sessile, margins entire, veins white below, bases rounded to heart-shaped, tips pointed; leaves reduced upward to leaflike bracts; 1 to 4 inches long.

Flowers: In clusters, 3-to-9-flowered; sepals 4 to 6, pink; petals 6, oblong-obovate, wrinkly; stamens 6 to 12; stigma green, club-like; involucral bracts small, hairy. June through August.

Fruits: Capsules 2-valved; seeds many.

Distribution: Wet ground, roadside ditches, waterways. Uncommon. Being eradicated because of its invasive tendencies.

In Kentucky: (AP-rare), IP, ME. Listed as a Severe Threat by the Kentucky Exotic Pest Plant Council.

Considered to be one of the worst invasive plants in the northern United States, this species had spread into the wetlands of glaciated North America by the late nineteenth and twentieth centuries and has become an aggressive plant in some areas, particularly in the Great Lakes Region, the St. Lawrence River Valley, and Hudson River Valley. Today, it is found growing in forty-two of the fifty states.

Purple loosestrife was introduced into the United States from Europe in the early 1800s in ship ballasts and as an ornamental plant. The first report of this species in North America was in Pursh's *Flora Americae Septentrionalis* in 1814. Two well known botanists, John Torrey and Asa Gray, were skeptical of the previous records, and Gray wrote in 1865 that its standing as a native plant was "not clear." Finally, in 1890, the sixth edition of *Gray's Manual of Botany* was

Invasive plant
Cher, Sen, Iroq, Shaw

published, in which the species was considered a plant naturalized from Europe and therefore not native to the North American flora.

When mature, this species is capable of having thirty to fifty stems from a single rootstock as well as producing approximately two to three million seeds per year. In Kentucky, purple loosestrife is found in the Purchase Area and in counties along the Ohio and Red rivers. It is uncommon in Jefferson County, but a few small colonies have been located along Beargrass Creek, Breckenridge Spring, in wetlands at Iroquois, and along the Ohio River.

on slender stalk, soon fading and hanging downward.

Origin: Native.

Life form: Perennial herb from fibrous roots.

Stems: Erect, single or with a few spreading, 4-angled branches, bright green to purple; 1 to 2 feet tall.

Leaves: Blades ovate, dull green, bases rounded, tips long-tapering, to 6 inches long; lower leaves similar, long-stalked, uppermost reduced, sessile.

Flowers: Small, in long terminal and upper axillary clusters; sepals tubular, 2-lipped; petals 5, fused, 2-lipped, upper lip notched, lower 3-lobed, larger; stamens 4; style 1; bracts at base of leaf stalk tiny. June through August.

Fruits: Achene small, enclosed by 3-ribbed, reflexed sepals that press against the flower stalk.

Distribution: Woods, woodland edges. Rare at Cherokee; uncommon at Iroquois.

In Kentucky: AP, IP, ME.

Lopseed
Lopseed Family
Phryma leptostachya L.
Phrymaceae

Key features: Leaves opposite, coarsely toothed; flowers pinkish purple to white, opposite, horizontally spreading

This unusual wildflower belongs to a family with only one genus, *Phryma,* which grows in eastern North America and northeastern Asia. The flowers, when in bloom, are held horizontally on the long slender stem, but when the petals fall off, they hang or "lop" downward against the stem—hence the common name lopseed.

Pensylvania smartweed

Smartweed Family
Polygonum pensylvanicum (L.) Small
Polygonaceae

Key features: Plants with jointed stems; flowers pink or purplish, occasionally white, in dense slightly nodding clusters; cylindrical sheath at base of leaf stalk hairless, papery.

Origin: Native.

Life form: Summer annual from a taproot.

Stems: Ascending to erect, branching, green or reddish green, smooth to hairy-glandular, swollen at the nodes; to 3 feet tall.

Leaves: Blades lanceolate to elliptic, margins entire, bases tapering, tips pointed, upper surface often with reddish purple blotch; 2 to 5 inches long.

Flowers: Many, small, in terminal and axillary clusters; sepals pink, 5-parted; petals absent; styles 2 or 4. June through September.

Fruits: Tiny, black, broadly oval, flattened on 1 side, concave on other.

Distribution: Disturbed ground in moist open woods, thickets, roadside ditches, pond margins, waterways, meadows. Common.

In Kentucky: AP, IP, ME.

Similar species: Spotted lady's thumb (*Polygonum persicaria* L.) is a summer annual from Europe and grows to 3 feet tall. The alternate leaves are lanceolate to elliptic, often with a reddish purple blotch on the upper surface. The **papery cylindrical sheath at the base of the leaf stalk has bristlelike hairs on the margin.** Many small, **pink to deeper rose flowers** are in slender terminal and axillary clusters. Cher, Sen, Iroq, Shaw, Chick. Disturbed moist ground in open woods, roadside ditches, thickets, waterways, pond margins. Common. June through September. In Kentucky: AP, IP, ME. Listed as a Significant Threat by the Kentucky Exotic Pest Plant Council.

Some of the species in this large genus are sharp or peppery in taste and their juices smart, or sting, the mucous membranes of the eyes, nose, and mouth—hence the name smartweed.

Soaking the plant in vinegar and wrapping it around one's head is an old folk remedy believed to cure a headache.

The seeds of spotted lady's thumb have been eaten by people since prehistoric times and are said to be similar to rice and maize in nutritional value. The foliage is high in amino acids and has been fed to horses and cattle in some parts of the world.

Smooth false foxglove

Figwort Family
Agalinis purpurea (L.) Pennell
Scrophulariaceae

Key features: Leaves opposite, narrow; flowers bell-shaped with 5 rounded lobes, pinkish purple with dark purple dots and yellow lines on the inside of throat.

Origin: Native.

Life form: Annual herb from fibrous roots.

Stems: Ascending to sprawling, ridged, diffusely or much branched; 1 to 3 feet tall.

Leaves: Blades linear, margins entire, midvein distinct, often produces smaller secondary branches from axils of primary leaves; 1 to 2 inches long.

Flowers: On ascending branches; sepals tubular, 5-lobed, small, lacking netlike veins; petals 5, lobes flaring; stamens 4; pistil 1; ¾ to 1½ inches wide. July through September.

Fruits: Capsules round, ¼ inch long; seeds numerous.

Distribution: Moist to wet ground. Uncommon.

In Kentucky: AP, IP, ME.

Similar species: **Common false foxglove** [*Agalinis tenuifolia* (Vahl.) Raf.] is a wiry stemmed plant that grows to 20 inches tall. The opposite **leaves are threadlike, mostly curled,** and to 1 inch long. Delicate purplish pink flowers, ½ inch long, have 5 lobes with the **upper lobes arching over the yellow fuzzy stamens and style.** Dry ground in open woods, woodland edges at the summit of Iroquois. Uncommon. August through October. In Kentucky: AP, IP, ME. (*syn=Agalinis besseyana* Britton)

The genus *Agalinis* contains several species that are considered to be parasitic on grass roots.

Eastern figwort

Figwort Family
Scrophularia marilandica L.
Scrophulariaceae

Key features: Plant tall, slender; leaves opposite, stalked, reduced upward; flowers small, tubular, reddish brown on outside, green inside, in loosely branched clusters.

Origin: Native.

Life form: Perennial from rhizomes.

Stems: Erect, angles rounded, sides grooved, occasionally branched, somewhat glandular-hairy; 5 to 7 feet tall.

Leaves: Blades ovate to lanceolate, usually hairy beneath, margins sharply toothed, bases round to heart-shaped, tips long-pointed; to 8 inches long.

Flowers: Sepals 5, green, deeply divided; petals irregularly 2-lipped, upper 2 lobes hooded, side lobes small, lower lobe reflexed; stamens 5, reddish brown; $1/4$ inch long. July through September.

Fruits: Capsule round; seeds wrinkled.

Distribution: Open woods, thickets. Uncommon.

In Kentucky: (AP-rare), IP, ME.

According to the Doctrine of Signatures, the fleshy knobs on the rhizomes were said to cure *scrofula,* often known as "tuberculosis of the neck," caused by an infection of the lymph nodes and to remove fig-warts.

Common periwinkle

Dogbane Family
Vinca minor L.
Apocynaceae

Key features: Evergreen woody vine; leaves opposite, leathery, with whitish green veins; flowers blue-violet, solitary on stalks in the upper leaf axils.

Origin: Europe and western Asia.

Life form: Perennial vine from rhizomes.

Stems: Trailing, shiny green, becoming woody with age.

Leaves: Blades ovate to elliptic, dark green above, lighter green below, margins entire, veins distinct, bases rounded to heart-shaped, tips pointed or blunt; 1 to 2 inches long.

Flowers: Sepals 5, small, margins entire; petals 5; stamens 5, anthers joined to each other; pistils 2; 1 inch wide. April through May.

Fruits: Pods, in pairs, slender, to 2 inches long; seeds several, tiny, black with a crown of hairs.

Distribution: Moist woods, waterways, shady roadsides, turf. Locally common in some areas of parks.

In Kentucky: AP, IP, ME. Listed as a Significant Threat by the Kentucky Exotic Pest Plant Council.

This popular ground cover belongs in a plant family that contains many important ornamentals as well as drug plants, alkaloids, and latex. In some tropical countries, poisoned arrows are made with an extract from species within this genus.

According to folklore, criminals on the way to the gallows in medieval Britain wore garlands of periwinkle around their necks.

Although this species can be found escaping into woods from old home sites and urban gardens, it mainly spreads by means of underground rhizomes and rootlets produced at the nodes. Forming dense colonies, the evergreen foliage smothers out native plants and has little ecological value to the local fauna. Another common name is running-myrtle.

Wild comfrey

Borage Family
Cynoglossum virginianum L.
Boraginaceae

Key features: Plant densely hairy; leaves basal and alternate above; flowers pale blue, small, borne in loose terminal clusters on long stalks.

Origin: Native.

Life form: Perennial herb from a taproot.

Stems: Erect, light green, rough with dense spreading hairs; 12 to 30 inches tall.

Leaves: Mostly basal; lower large, elliptic, base tapering into a stalk, margins entire; alternate stem ones few, smaller, leaf bases heart-shaped, clasping, tips blunt to short-pointed; all densely hairy; to 10 inches long.

Flowers: Few-flowered; sepals 5 green, deeply lobed, hairy; petals 5, funnel-shaped, lobes rounded; stamens 5; pistil 1; less than ¼ inch wide. April through June.

Fruits: Nutlets 4, together, small, rounded, covered with soft prickles.

Distribution: Moist low woods. Uncommon.

In Kentucky: AP, IP, ME.

Other common names for this graceful wildflower are dog-tongue and hound's-tongue, both referring to the shape and rough texture of the leaves, which look like a dog's tongue. The generic name is derived from Greek *cynos* "of the dog" and *glossa* meaning "tongue."

In traditional medicine a poultice made from the large leaves was used to relieve insect bites. The leaves beaten into small pieces and added to old swine grease had the power to heal dog bites, and people put the leaves under their feet to keep dogs from barking at them.

Some persons are allergic to the downy leaves, which can cause dermatitis.

Virginia bluebells

Borage Family
Mertensia virginica (L.) Pers. ex Link.
Boraginaceae

Key features: Plant fleshy; leaves alternate, pale green with whitish cast; flowers blue, buds pink, nodding in showy, 1-sided terminal clusters.

Origin: Native.

Life form: Perennial herb from fibrous roots.

Stems: Erect or ascending, simple or branching, smooth; 1 to 2½ feet tall.

Leaves: Blades elliptic to obovate, smooth, margins entire, bases tapering into a long stalk; upper ones similar,
reduced, short-stalked to clasping; up to 8 inches long.

Flowers: Sepals green, cuplike, divided into 5 blunt lobes; petals 5, trumpet-shaped with outward-flaring lobes; stamens 5, white, anthers light brown; style white, long, slender; ¾ to 1 inch long. April through May.

Fruits: Nutlets 4, ovoid.

Distribution: Moist woods, woodland edges, along creek banks. Uncommon.

In Kentucky: AP, IP, ME.

This beautiful wildflower is also called Roanoke bells and Virginia cowslip, both referring to the colony in Virginia. Another name, inspired by garden writers of the eighteenth century, was "Jefferson's blue funnel flowers," as it was a popular plant grown in Thomas Jefferson's famous Monticello gardens.

The deep pink bud opens into a rich sky-blue flower, rarely white or pink, in early spring and attracts such pollinators as hummingbirds, skippers, and butter-flies. After flowering, the foliage turns yellow, dies back to the ground, and disappears.

Mertensia was named after Franz Karl Mertens, a prominent eighteenth-century German botanist.

Honesty
Mustard Family
Lunaria annua L.
Brassicaceae

Key features: Plant from a deep taproot; leaves heart-shaped to triangular; flowers purple with darker purple veins; fruit pods nickel-like, turning papery with age.

Origin: Europe.

Life form: Annual or biennial herb.

Stems: Stout, erect, ridged, hairs white; 1 to 2 feet tall.

Leaves: Alternate above, opposite below; blades rough above, hairs white, margins coarsely to bluntly toothed, tips short- to long-pointed; upper stem leaves reduced, irregularly toothed, sessile; to 5 inches long.

Flowers: Several on stalks ½ to 1 inch long, in terminal branching clusters; sepals 4, greenish purple; petals 4; stigmas minute; ¾ to 1 inch wide. April through June.

Fruits: Pods very thin, flat, net-veined, turning whitish tan and papery with age, 1 to 2 inches long and wide; seeds dark, flat, kidney-shaped, winged.

Distribution: A small population has taken hold in Iroquois southwest of stables in open woods; rare along Beargrass Creek in Cherokee and Seneca. First sited in Seneca in 2006.

In Kentucky: (IP-previously only collected in 2 counties: Campbell, Bracken; a new record for Jefferson County.)

This plant was introduced into the United States from Europe as an ornamental because of the attractive bright purple flowers and unique dried fruit. It has escaped from cultivation into open woods and roadsides, where it is becoming established in several states.

Also called money plant, moonwort, satin flower, and silver dollar plant, the seedpods resemble thin, papery coins and are valued for dried bouquets.

The generic name *Lunaria* means "moon" and refers to the round silvery fruits.

Wild hyacinth

Grape-Hyacinth Family
Camassia scilloides (Raf.) Cory
Hyacinthaceae

Key features: Plant from a bulb; leaves basal, narrow; flowers blue to whitish

blue, 6 stamens prominent, produced in loose terminal clusters.

Origin: Native.

Life form: Perennial herb.

Stems: Lacking except for the flowering stalks to 2 feet tall.

Leaves: Blades linear, shorter than the flower stalk, veins parallel, tips pointed; 8 to 18 inches tall.

Flowers: Tepals 6, linear-oblong, spreading; stamens 6, anthers yellow; stigma 3-lobed; involucral bract below each flower linear, falling early; 1 inch wide. April through May.

Fruits: Capsule 3-angled; seeds many, small, black, shiny.

Distribution: Shady, moist woods, limestone outcrops. Seems to be coming back on limestone ledges where invasive shrubs have been removed. Uncommon, but locally common on ledges above archery range.

In Kentucky: (AP-rare), IP, (ME-rare).

The generic name comes from the Native American word *quamash* or *camass* and *Hyacinth,* a pre-Greek word that refers to the color of the sea.

The common name Eastern camass is often used for wild hyacinth and can be confused with the highly poisonous species belonging to the genus *Zigadenus,* known as death-camass.

Siberian squill

Grape-Hyacinth Family
Scilla siberica Haw.
Hyacinthaceae

Key features: Plant from a bulb; leaves basal, shiny green; flowers brilliant blue, with darker blue midveins, saucer-shaped.

Origin: Eurasia.

Life form: Perennial herb.

Stems: Lacking except for the purplish green flowering stalks to 6 inches tall.

Leaves: Two to 7, narrow, smooth, margins entire, tips tapering; 3 to 5 inches long.

Flowers: One to 3, nodding; tepals 6, blue, oblong-linear; stamens 6, anthers blue; pistil green; ½ to 1 inch wide; single leaflike bract below each flower. March through May.

Fruits: Capsules round, bumpy, green turning brown; seeds several, tiny.

Distribution: Disturbed shady woods at Canebrake Management Area at Seneca, where it is invasive; rare in Cherokee.

In Kentucky: Naturalized populations have been documented from Campbell and Jefferson Counties.

This striking ornamental herb is also called scilla and is used in naturalistic settings, often blooming in early spring when snow is still on the ground. In deciduous woods, shaded lawns, and other favorable settings, this species can spread quickly, forming large colonies over several years.

Native to the temperate regions of the Old World, there are close to 100 species in this genus, with many different horticultural varieties. Scilla has been cultivated since 1796.

Biennial waterleaf

Waterleaf Family
Hydrophyllum appendiculatum Michx.
Hydrophyllaceae

Key features: Leaves alternate, palmately lobed, often mottled grayish blue; flowers purple with 5 tiny, reflexed appendages alternating with the sepals.

Origin: Native.

Life form: Biennial herb from a taproot.

Stems: Upright, branched above, hairy; up to 2 feet tall.

Leaves: Long-stalked, blades palmate with 5 to 7 broad, irregularly lobed or cleft segments about as long as wide, dark green above or sometimes mottled, margins toothed; to 6 inches long.

Flowers: In loose clusters that rise above the leaves; sepals 5, narrow, long-tapering, densely silky hairy; petals 5, bell-shaped; stamens 5, anthers brown, extending beyond the petals or equal to; pistil 1; l inch wide. April through June.

Fruits: Capsule round; seeds 1 to 3, tan, wrinkled.

Distribution: Moist woods, often forming extensive colonies along Beargrass Creek. Common at Cherokee; rare at Shawnee.

In Kentucky: (AP-rare), IP, (ME-rare).

Similar species: Hairy waterleaf (*Hydrophyllum macrophyllum* Nuttall) is a perennial herb with **leaves deeply pinnately divided into 7 to 13 lobes nearly to the midvein.** The blades, 3 to 8 inches long, are green and often **mottled in shades of light and dark green. Flowers are dull white** with stamens and styles protruding beyond the petals. Iroq. Moist woods. Rare. May through June. In Kentucky: (AP-rare), IP, (ME-rare).

The Waterleaf Family is small but widely distributed throughout the world. Because of the attractive flowers and foliage, several species are cultivated as garden ornamentals in such genera as *Nemophila, Wigandia,* and *Phacilia.*

Biennial waterleaf is also called appendaged waterleaf, and this refers to the five tiny reflexed appendages that alternate with the sepals.

Miami-mist

Waterleaf Family
Phacelia purshii Buckley
Hydrophyllaceae

Key features: Leaves alternate, pinnately lobed; flowers saucer-shaped, petals 5-lobed, lavender blue, fringed, center white.

Origin: Native.

Life form: Summer annual from fibrous roots.

Stems: Erect or lax, multiple from base, angled, hairy; to 16 inches tall.

Leaves: Stalked below, sessile to clasping above; oblong in general outline, blades 5-to-9-lobed, oblong or lanceolate, margins entire, lobes short-pointed; 1 to 2 inches long.

Flowers: Many in coiled 1-sided terminal or axillary clusters, elongating in fruit; sepals 5, green, linear; petals 5-lobed; stamens 5, exserted, filaments white, anthers bluish purple, hairy; style white, 2-forked; ¾ inch wide. April through June.

Fruits: Capsule round, 2-valved; seeds many, angular.

Distribution: Moist thickets, waterways, low moist woods, especially in alluvial soils. Uncommon; locally common at Seneca.

In Kentucky: AP, (IP-mostly Bluegrass region), (ME-rare).

Phacelia is from the Greek *phacelos* "cluster or fascicle" and refers to the crowded flowers. The species, *purshii,* is named for its discoverer, Frederick Traugott Pursh.

This beautiful flower is so abundant in some areas in Kentucky that it is considered an obnoxious weed, establishing itself in bare ground along waterways.

The Cherokee Indians made a poultice that was used for swollen joints.

Leaves: Basal, often crowded; blades lanceolate, slightly curving or arching, elongating to 7 inches long; spathe leaves green, ovate-lanceolate, enclosing the flowers.

Flowers: One-to-3-flowered; sepals ("falls") 3, petal-like, spreading and curving downward; petals 3 ("standards"), purple, oblanceolate, spreading and erect; 2 to 3 inches wide. April through May.

Fruits: Capsule ovoid to ellipsoid, sharply 3-angled; seeds orange-brown.

Distribution: Moist shady woods, ravine slopes. Uncommon, but can be locally common.

In Kentucky: AP, (IP-rare in northern section), ME.

Dwarf crested iris

Iris Family
Iris cristata Soland. ex Aiton
Iridaceae

Key features: Plant poisonous; leaves basal, swordlike, light green with whitish tinge; flowers showy, purple, with a yellow or white crest on the downward-curving sepals.

Origin: Native.

Life form: Perennial herb with creeping or horizontal rootstocks.

Stems: Lacking except for flower stalks 4 to 6 inches tall.

This beautiful spring wildflower often produces patches of sterile shoots with only leaves and no flowers; such patches can be seen at Iroquois.

The family is valued principally for its ornamental plants such as crocus, gladiolus, freesia, and croscosmia. The genus name is from the mythological goddess of the rainbow and refers to the flower colors within the group.

The Cherokee Indians would pulverize the roots in hog's lard, sheep suet, and beeswax to make a salve for ulcers. The roots are said to be sweet at first but change quickly to a burning sensation, stronger than *Capsicum* (hot chili peppers).

Ground-ivy

Mint Family
Glechoma hederacea L.
Lamiaceae

Key features: Stems 4-angled, creeping; leaves opposite, stalked, rounded; flowers 1 to 3, purplish pink to blue, tubular, borne in the leaf axils.

Origin: Europe.

Life form: Perennial herb from fibrous roots and rhizomes.

Stems: Erect to mostly creeping and rooting at the nodes, smooth; to 3 feet long.

Leaves: Blades round to kidney-shaped, green to purplish green, palmately veined, margins bluntly toothed; up to 1 inch wide.

Flowers: Sepals 5, tubular; petals 5, 2-lipped, upper lip 2-cleft, arched, lower lip 3-lobed; stamens 4; pistil 1; ½ inch long. April through June.

Fruits: Nutlets ovoid, dark brown, in groups of 4 (or fewer by abortion) each with 1 seed.

Distribution: Disturbed ground, especially invasive in moist shady woods, waterways, turf, fields, roadsides, cultivated beds.

In Kentucky: AP, IP, ME. Listed as a Severe Threat by the Kentucky Exotic Pest Plant Council.

Ground-ivy, also called alehoof, has been associated with the ale industry for hundreds of years. It was the most widely used seasoning in brewing ale until the German discovery of the value of hops. Ground-ivy was believed to flavor and preserve the ale as well as help it clear.

Another name is gill-over-the-ground. The word *gill* is from the French word that means "fermented ale." The common names creeping Charlie, roving Charlie, and run-away-Nell describe the creeping nature of this rampant weed, which is capable of climbing up lower tree trunks.

Wild sage

Mint Family
Salvia lyrata L.
Lamiaceae

Key features: Leaves mostly basal, pinnately lobed or lyre-shaped; flowers blue to violet, tubular, in interrupted whorls along upper part of stem.

Origin: Native.

Life form: Perennial herb from fibrous roots.

Stems: Erect, simple, occasionally branching, 4-angled, hairs long; to 15 inches tall.

Leaves: Blades oblong to ovate in general outline, margins irregularly toothed, wavy to pinnately lobed into rounded segments, terminal lobe largest, central vein reddish, hairs long; stem leaves in pairs, 1 to 3, blades lanceolate, reduced, short-stalked to clasping; 3 to 6 inches long.

Flowers: Sepals 2-lipped, upper lip purple, entire or with 3 sharp-pointed teeth, lower lip 2-lobed, longer, green with 2 sharp-pointed teeth; petals 2-lipped, upper lip smaller than lower; stamens 2, anthers brownish purple; pistil 1, style exserted. April through June.

Fruits: Nutlets 4, dull dark brown.

Distribution: Disturbed ground in fields, rocky woods, waterways, roadsides, turf. Common.

In Kentucky: AP, IP, ME.

Other common names, such as cancer root and cancerweed, refer to the plant's prolific growth habit, like a "cancer" upon the earth. Folk medicine thought that such a plant should be able to cure any cancer disease. The leaves were said to remove warts, and a salve made from the roots was applied to sores.

Lyreleaf sage describes the lyre-shaped leaves.

Violet wood-sorrel
Wood-sorrel Family
Oxalis violacea L.
Oxalidaceae

Key features: Plant from stolons, often forming colonies; leaflets 3, heart-shaped, purplish red below; flowers purplish pink in loose clusters arising above the leaves.

Origin: Native.

Life form: Perennial herb.

Leaves: Basal, long-stalked; blades grayish green to purplish green above, sometimes with brownish blotches, margins entire, tips notched; to 1 inch wide.

Flowers: Few, on delicate stalks to 7 inches long; sepals 5, green, tipped with orange gland; petals 5, joined at base, flaring outward above; stamens 10, anthers yellow; styles 5; to ¾ inch wide. April through May.

Fruits: Capsules slender, splitting into 5 sections; seeds tiny, light brown.

Distribution: Rocky open woods, limestone ledges. At Cherokee, this species has been found in a few new locations since the removal of bush honeysuckle. Uncommon.

In Kentucky: AP, IP, ME.

The wood-sorrels are often called sourgrass and contain oxalic acid, which is secreted as calcium oxalate crystals. When ingested in large quantities, these crystals can be harmful to the kidneys.

The leaflets of *Oxalis* open up during the day and fold downward at night and in cold weather.

Forest phlox
Phlox Family
Phlox divaricata L.
Polemoniaceae

Key features: Leaves opposite, narrow, sessile, widely spaced on stem; flowers pale blue to lavender, petals notched at the tip, in loose terminal clusters.

Origin: Native.

Life form: Perennial herb from a taproot and aboveground stolons.

Stems: Flowering stems erect, greenish red, hairy or smooth; vegetative stems on ground; 12 to 20 inches tall.

Leaves: Fertile shoots: opposite, sessile, blades lance-ovate, margins entire, tips blunt; basal shoots: opposite blades elliptic to oblong, sessile or clasping; to 3 inches long.

Flowers: Sepals hairy, with 5 deeply divided narrow, slightly flaring lobes; petals 5-lobed; stamens 5, hidden inside tube; about 1 inch wide. April through May.

Fruits: Capsules rounded, 3-valved, enclosed by the hairy sepals.

Distribution: Moist woods. Rare.

In Kentucky: AP, IP.

This family, which contains many popular garden ornamentals, is small but exhibits great diversity of habit, ranging from trees and lianas to small annual and perennial herbs. The pollination mechanisms also vary greatly within the species and attract bees, flies, beetles, butterflies, moths, hummingbirds, and bats in some tropical genera.

It is sometimes called Sweet William, but this name is more accurately applied to the cultivated *Dianthus barbatus* in the Pink Family.

Jacob's-ladder
Phlox Family
Polemonium reptans L.
Polemoniaceae

Key features: Leaves pinnately compound, leaflets opposite, sessile; flowers deep blue to lavender, bell-shaped, in loose drooping clusters in the upper leaf axils.

Origin: Native.

Life form: Perennial herb from fibrous roots.

Stems: Slender, erect or weak and spreading, branched above, green to reddish green, smooth; 10 to 16 inches tall.

Leaves: Basal, long-stalked, upper sessile; blades ovate to lanceolate, glossy green, margins entire, bases tapered, tips rounded to short-pointed; 1/2 to 2 inches long.

Flowers: Sepals 5, green, ovate; petals 5; stamens 5, anthers white, shorter than petals; style 3-forked at tip; 1/2 to 1 inch wide. April through May.

Fruits: Capsule round, 3-valved, surrounded by the sepals that enlarge after flowering.

Distribution: Moist woods. Uncommon.

In Kentucky: (AP-rare), IP, ME.

This beautiful spring wildflower has leaflets that are arranged like the steps of a ladder—hence the common name. The genus honors Polemon, an early Athenian philosopher.

Native Americans made an infusion from the roots that was used to cure coughs, colds, and lung problems.

Dwarf larkspur

Buttercup Family
Delphinium tricorne Michx.
Ranunculaceae

Key features: Plant poisonous; leaves mostly basal below, alternate above, palmately 5-parted; flowers white and purple, spurred, in loose terminal clusters.

Origin: Native.

Life form: Perennial herb from a cluster of short, tuberous roots.

Stems: Erect, simple, succulent, purplish green; to 2 feet tall.

Leaves: Palmate, lobes many, wedge-shaped to linear-oblong each with 2 to 3 secondary lobes, margins entire, tips pointed; 2 to 3½ inches wide.

Flowers: Sepals 5, petal-like, purple, bluish white to pinkish, spreading outwards; petals 4, upper 2 small, white and extend backwards into a nectar spur, lower 2, purple, hairy; 1 to 1½ inches wide. April through May.

Fruits: Pods 3, brown, spreading slightly outward; seeds 3-angled, black, smooth.

Distribution: Moist to dry woods, limestone ledges. Uncommon.

In Kentucky: AP, IP, ME.

This family has many weedy members in addition to those that are used extensively in ornamental gardens, such as monkshood, hellebores, winter aconite, buttercups, and love-in-the-mist. A number of genera are extremely poisonous and have caused deaths in both livestock and humans.

Sharp-lobed hepatica
Buttercup Family
Hepatica acutiloba DC.
Ranunculaceae

Key features: Leaves and flowering stalks soft-hairy; leaves broadly 3-lobed, glossy green turning bronze in winter; flowers solitary, blue, lavender, pink, or white.

Origin: Native.

Life form: Perennial herb from rhizomes.

Stems: Lacking except for the flower stalks 5 to 10 inches tall.

Leaves: Few to many on stalks to 6 inches long, from basal tufts that persist through winter; blades broadly ovate, 3-lobed, margins entire, bases unequally heart-shaped, tips pointed; 2 to 3 inches long and wide.

Flowers: One to many; outer bracts 3, green, about equaling the sepals; sepals 5 to 12, petal-like; petals absent; stamens white, numerous; pistils numerous; ¾ to 1 inch wide. March through April.

Fruits: Achenes lance-ovoid.

Distribution: Moist wooded slopes along creek banks in calcareous soils. Rare.

In Kentucky: AP, IP.

The fragile blossoms of sharp-lobed hepatica herald the coming of spring. The downy buds are often hidden by the old bronzed and weather-beaten leaves of last year, while the new, fresh, hairy leaves appear after the short-lived flowers die.

An example of the Doctrine of Signatures (a superstition which held that every medicinal plant had an outward sign or "signature" that revealed its specific use), this plant was used in ancient times to remedy liver ailments. The generic name is from *hepaticus,* Latin for "liver," and another common name, liverleaf, refers to the resemblance of the leaf to that of a three-lobed mammalian liver.

Spring bluets

Madder Family
Houstonia caerulea L.
Rubiaceae

Key features: Plant delicate; leaves mostly basal, upper opposite, few, reduced; flowers solitary, pale blue to violet with yellow center.

Origin: Native.

Life form: Perennial herb from short, thin rhizomes.

Stems: Solitary, erect, sparingly branched below; to 6 inches tall.

Leaves: Basal rosette with blades oblong-spatulate to elliptic, margins entire; ¼ to ¾ inch long; upper ones clasping.

Flowers: Produced at the tips and upper leaf axils on threadlike stalks; sepals 4, green, tiny, narrow; petals salverform, 4-lobed; stamens 4; pistil 1, stigma 2; ½ inch wide. March through June.

Fruits: Capsule flattened; seeds 4 to 20, pitted.

Distribution: Moist to dry open woods, mossy banks, fields, roadsides. Common in Iroquois; rare at Cherokee.

In Kentucky: AP, IP, ME.

Similar species: Small bluet (*Houstonia pusilla* Schoepf) is an annual herb to 2½ inches tall. Borne at the tip of a threadlike stalk is a solitary flower with 4 green sepals and **4 blue petals with a reddish purple center.** Uncommon in Seneca; rare in Cherokee. In thin soil over limestone outcrops, mown areas. March through May. In Kentucky: (AP-rare), ME. First record for Jefferson County.

Spring bluets, also known as angel-eyes, Quaker ladies, Quaker bonnet, and innocence, was said to be much admired by the early settlers. The Cherokee Indians made a tea to stop bed wetting.

According to herbarium records, small bluet was absent from the Bluegrass until it was recently collected from Seneca Park. It is easily overlooked because of its small size; it is barely visible growing in its preferred habitats in mown areas. The delicate flowers are beautiful when viewed up close.

The genus was named in honor of the Scottish physician and botanist Dr. William Houston, who collected plants in Mexico and the Caribbean.

Birdseye speedwell

Figwort Family
Veronica persica Poir.
Scrophulariaceae

Key features: Stems low-growing; leaves alternate above, opposite below, toothed; flowers small, pale blue, with deeper blue lines on slightly curved stalks, axillary.

Origin: Asia.

Life form: Summer annual from fibrous roots.

Stems: Ascending to lying on the ground, often rooting at the lower nodes and forming dense mats; 4 to 12 inches long.

Leaves: Blade ovate to rounded, margins toothed or lobed, short-stalked to clasping; less than 1/2 inch long.

Flowers: Solitary, on stalks from the upper leaf axils; sepals 4; petals 4; stamens 2; pistil 1; about 1/3 inch wide. March through June and sporadically during the warm winter months.

Fruits: Capsules heart-shaped, notched, wider than long, surrounded by 4 persistent sepals.

Distribution: Disturbed ground, especially in turf, cultivated beds. Common.

In Kentucky: AP, IP, ME.

Similar species: Purslane speedwell (*Veronica peregrina* L.) is a native annual herb with **erect stems to 15 inches tall.** The sessile to short-stalked smooth leaves are oblong to linear, mostly blunt-tipped, with the lowest ones opposite and the upper ones alternate. **Tiny, solitary, white flowers in distinct racemes or spikes** cover the upper 1/2 to 3/4 of the stem with reduced leaves. The fruit capsule is heart-shaped with a very short style, the sepals slightly longer than the capsule. Cher, Sen, Iroq, Shaw, Chick. Disturbed open woods, waterways, roadsides, cultivated beds. Common. April through September. In Kentucky: AP, IP, ME.

The genus name *Veronica* means "true image" and was named for St. Veronica. An early Christian legend depicts St. Veronica accompanying Christ on the trip to Calvary. Wiping his face with her handkerchief, she notices a miraculous true image of his features in the cloth.

Thyme-leaved speedwell

Figwort Family
Veronica serpyllifolia L.
Scrophulariaceae

Key features: Stems often mat-forming; leaves opposite below, alternate above; flowers small, bluish white with darker blue lines, in terminal racemes that soon elongate.

Origin: Europe.

Life form: Perennial herb from fibrous roots.

Stems: Ascending to creeping, rooting at the nodes; to 10 inches long.

Leaves: Lower short-stalked to clasping; blades ovate to oblong, margins entire to slightly blunt toothed; upper leaves alternate, narrow, clasping, reduced, uppermost becoming bracts; ¹/₂ inch long or less.

Flowers: Sepals 4, deeply lobed; petals 4; stamens 2, exserted; pistil 1, style long. April through July.

Fruits: Capsule heart-shaped, broader than long, flat; seeds tiny, orange, granular.

Distribution: Disturbed ground in turf, open moist woods, roadsides, fields. Common.

In Kentucky: AP, IP, (ME-rare).

Similar species: Common speedwell (*Veronica officinalis* L.) is a creeping perennial herb from Europe that roots at the nodes, often forming dense mats. The **leaves are all opposite** and are oval to obovate with finely toothed margins. **Flowers are produced in axillary racemes** and are small, **bluish white, sometimes with darker blue lines.** Cher, Sen, Iroq, Shaw, Chick. Disturbed ground in turf, open moist woods, roadsides, waterways. Uncommon. April through July. In Kentucky: AP, IP.

Native Americans made a juice and a tea from the leaves of the thyme-leaved speedwell that was used for earaches, chills, and coughs.

Common blue violet

Violet Family
Viola sororia Willd. var. *sororia*
Violaceae
(*syn=Viola papilionacea* Pursh.)

Key features: Plant stemless; leaves basal, long-stalked, heart-shaped; flowers deep violet, lavender to white, single on long stalks held above the ground.

Origin: Native.

Life form: Perennial herb from a thick, creeping rhizome.

Stems: None except for the flowering stalk to 3 to 8 inches tall.

Leaves: Blades dark green, upper and lower surfaces smooth to slightly hairy, margins bluntly to sharply toothed, bases heart-shaped, tips rounded or short-pointed; 1 to 2 (5) inches wide; leaflike stipules at base of leaf stalk narrow, margins entire, tips sharp-pointed.

Flowers: Single, ¾ to 1 inch wide, arising from the base of the plant; sepals 5, green ovate-lanceolate; petals 5, unequal, upper petals oblong, side petals bearded, lower spurred with dark lines leading to a yellow base; stamens 5, fused; spring flowers exceed leaves, later flowers often shorter than leaves. March through May and sporadic in the fall.

Fruits: Capsule 3-angled, smooth, mottled purple and brown; seeds tan, mottled with brown and gold, often expelled explosively.

Distribution: Moist to dry ground in disturbed woods, waterways, turf, fields, cultivated beds. Abundant.

In Kentucky: AP, IP, ME.

This native lawn weed is the state flower of Wisconsin, Illinois, New Jersey, and Rhode Island. The leaves are high in Vitamins A and C and can be used in raw salads or as cooked greens in spring. Nowadays, the flowers are candied and used to decorate cakes and other pastries.

Violets produce two kinds of flowers—those that are showy and produced in the spring, and those that are inconspicuous and lie on the ground in the summer. The latter are called cleistogamous. These flowers are small, budlike, and self-pollinating. The showy spring flowers with colored petals attract pollinating insects, which are guided by the dark lines. Nectar is secreted into the spur from the bases of the two lowermost stamens. As the insect touches the anthers, pollen is showered on its back.

American water-willow

Acanthus Family
Justicia americana (L.) Vahl.
Acanthaceae
(*syn=Dianthera americana* L.)

Key features: Plant growing in shallow water or mudflats, colonial; leaves opposite, willowlike; flowers pale purple or white with purplish brown markings on lower petals.

Origin: Native.

Life form: Perennial herb from thick, cordlike rhizomes.

Stems: Simple or branched, ridged, often reclining below and rooting at the nodes, ascending above to 3 feet.

Leaves: Narrow, sessile or short-stalked, tapered at the base, margins entire to slightly wavy, midvein distinct; 1½ to 5 inches long.

Flowers: Produced in clusters on long stalks from the upper leaf axils; sepals 5; petals 5, forming a 2-lipped corolla, the upper lip, notched, recurved, lower petals 3-parted, spreading. June through August.

Fruits: Capsule 2-valved; seeds brown, small, densely warty, nearly circular.

Distribution: Shallow water, mudflats, often forming extensive colonies in and along Beargrass Creek. Uncommon to locally common.

In Kentucky: AP, IP, ME.

This species spreads both by rhizomes and by seeds that are forcibly ejected from the fruiting capsules. Although this plant can form dense stands that interfere with recreational activity, it also reduces erosion along creek banks subject to wave action and flow. The dense vegetation provides a habitat for fish and invertebrates.

The genus honors James Justice, an eighteenth-century Scottish botanist and horticulturalist.

Limestone wild petunia

Acanthus Family
Ruellia strepens L.
Acanthaceae

Key features: Leaves opposite, bases tapering into a short, slender stalk; flowers lavender blue to pink, sepals narrow.

Origin: Native.

Life form: Perennial herb from a taproot.

Stems: Simple to sparingly branched, angular, smooth to slightly hairy; 1 to 3 feet tall.

Leaves: Blades ovate-lanceolate, margins slightly wavy to entire, tips pointed; 1 to 4 inches long.

Flowers: One to 3, on terminal and upper axillary stalks; sepals 5; petals funnel-form with 5 flaring lobes, darker purple lines within; 2 inches long. May through July.

Fruits: Capsule smooth, brown, usually overtopped by narrow sepals.

Disribution: Woodland edges, open rocky woods especially along Beargrass Creek. Uncommon.

In Kentucky: AP, IP.

Similar species: Carolina wild petunia [*Ruellia caroliniensis* (J.F. Gmel) Steud] is a native species that has 1 to 3 **purplish pink**, funnel-form flowers, 1 to 2 inches long, in the upper leaf axils. The hairy **sepals are very narrow to bristlelike**. Iroq. Moist to dry fields, woodland edges. Uncommon. May through July. In Kentucky: AP, IP, ME.

Both species are aptly named, as they resemble the popular garden petunia. Other ornamentals in this mostly tropical family are thunbergia, firecracker plant, and beloperone. Acanthus leaves are also the main motifs in classical Greek temples.

The genus honors the sixteenth-century French herbalist Jean Ruelle.

Chicory

Aster Family

Cichorium intybus L.

Asteraceae

Key features: Plant with milky latex; leaves entire, lobed, or toothed; flower heads sky blue, pink, or occasionally white, dandelion-like, with notched tips.

Origin: Eurasia.

Life form: Perennial herb from a deep taproot.

Stems: Erect, branched above, hollow, becoming woody and reddish with age; 1 to 3 feet tall.

Leaves: Basal rosette and lower ones long-stalked; blades oblong-lanceolate, deeply or shallowly lobed to toothed; upper ones alternate, smaller, margins entire or slightly toothed, bases clasping; to 5 inches long.

Flowers: Heads in terminal or axillary clusters, few-flowered, on short branches; ray florets only; anthers blue, surrounding the style, stigma blue; involucral bracts greenish, outer shorter, inner longer. June through October.

Fruits: Achenes light brown, mottled, obovate, 4-to-5-angled.

Distribution: Disturbed ground in fields, roadsides, thickets, woodland edges. Common when escapes mowing.

In Kentucky: AP, IP, ME. Listed as a Severe Threat by the Kentucky Exotic Pest Plant Council.

Considered a very old "companion" of humans, the plant bears the name chicory or some derivative of it throughout the world.

Chicory is native to Europe and Asia and has been grown as a coffee substitute, imparting distinctive flavor and color since the mid-eighteenth century, first in Italy and later in Germany.

It was first introduced into the United States in 1785 by Governor Bowdoin of Massachusetts, who considered it a valuable salad green. It has since spread throughout the United States.

The plant will ooze white milk when scratched. Because of its similarity in appearance to milk, in the mid-

Invasive plant
Cher, Sen, Iroq, Shaw, Chick

seventeenth century the British herbalist Nicholas Culpepper recommended chicory milk for nursing women pained by an abundance of milk. He also prescribed it for "sore eyes that are inflamed" because the beautiful blue flowers, like eyes, close at night "to sleep."

Canada thistle

Aster Family
Cirsium arvense (L.) Scop.
Asteraceae

Key features: Plant from extensive rhizomes, often forming large colonies; leaves alternate, irregularly lobed with strong spines on the margins; flower heads purplish pink, small.

Origin: Eurasian and North Africa.

Life form: Perennial herb.

Stems: Erect, slender, branched above, leafy, ridged; to 4 feet tall.

Leaves: Sessile, lower elliptic to oblanceolate, entire to deeply pinnately lobed, upper surface smooth, cobweb-like hairs below; upper leaves smaller, blades elliptic to oblong, margins entire.

Flowers: Heads rounded, numerous, male and female on separate plants; disk florets purplish pink, tubular. July through October.

Fruits: Achenes oblong, dark brown, flattened, slightly curved; tufts of white feathery hairs at apex.

Distribution: Disturbed ground along roadsides, fields, waterways, thickets, cultivated beds. Common.

In Kentucky: AP, IP. Listed as a Severe Threat by the Kentucky Exotic Pest Plant Council.

This thistle is a very serious weed throughout the world. It is found in the southern half of Canada and the northern half of the United States. It reached North America in the mid-1800s and quickly became established, being transported through contaminated farm seed brought over by both the English and French colonists. It was first collected in 1821 in Montreal, where it

Invasive plant
Cher, Sen, Iroq, Shaw, Chick

was already an established troublesome weed, but Vermont was the first state to pass legislation to control this weed, in 1795.

Large colonies of Canada thistle may consist of only male plants where no seed is produced. Female plants produce abundant seed, and a single plant can produce up to 1,500 seeds. The underground horizontal roots descend twelve to twenty-four inches below the surface and often go untouched by farm equipment. However, if the roots are cut, the segments can survive and produce new plants. It spreads mostly by vegetative means.

This thistle is the only one with male and female flowers produced on separate plants. It has been spreading since 2005 in all parks.

Mist-flower

Aster Family
Conoclinium coelestinum (L.) DC.
Asteraceae
(*syn=Eupatorium coelestinum* L.)

Key features: Leaves opposite, distinctly 3-veined; flower heads purplish blue to occasionally white, in broad, flat-topped terminal clusters.

Origin: Native.

Life form: Perennial herb from slender rhizomes.

Stems: Erect, short-hairy, occasionally branched; 1 to 3 feet tall.

Leaves: Blades triangular to ovate, margins bluntly to sharply toothed, bases squared or rounded, tips pointed; 1 to 3 inches long.

Flowers: Heads in clusters, 1 to 1½ inches wide; inner disk florets tubular with 5 purple lobes, style exserted, forked; involucral bracts several, long-pointed, hairy. July through October.

Fruits: Achenes tiny, grayish brown, pitted, 5-ribbed, tapered at the base, hairs tufted.

Distribution: Woodland edges, fields, pond margins, roadside ditches, thickets. Common.

In Kentucky: AP, IP, ME.

This attractive plant, also called wild ageratum, resembles the cultivated ageratum and is often used in perennial flower beds. It spreads easily because of the creeping underground rhizomes and can become weedy.

Coelestinum means "heavenly" and refers to the color of the flower.

Carolina elephant's-foot

Aster Family
Elephantopus carolinianus Raeusch
Asteraceae

Key features: Stems hairy; leaves basal and alternate, broad; flower heads small, bluish purple to white, in rounded, compact terminal clusters.

Origin: Native.

Life form: Perennial herb from fibrous roots.

Stems: Erect, branched above, green to purplish green; 1 to 3 feet tall.

Leaves: Lower blades oblanceolate, elliptic to spatulate, margins blunt to sharply toothed, heavily veined, bases abruptly narrowed into a winged stalk, tips pointed; upper ones reduced, sessile; 3 to 10 inches long.

Flowers: Disk florets with petals unequally lobed; stamens 5, anthers united around the exserted style; involucral bracts 3, triangular to heart-shaped, margins membranous, resinous, tips pointed. July through September.

Fruits: Achenes small, white, hairy with 5 bristles.

Distribution: Dry to moist ground in open woods, thickets, waterways, roadsides, fields. Common.

In Kentucky: AP, IP, ME.

The large basal leaves, which usually disappear when the flowers open, are said to resemble an elephant's foot—hence the common name. The genus name *Elephantopus* comes from the Greek—*elephas* "elephant" and *pous* "foot"—which also describes the shape of the leaves.

Woodland lettuce

Aster Family
Lactuca floridana (L.) Gaertn.
Asteraceae

Key features: Plant with milky sap; leaves alternate with winged stalks; flower heads numerous, small, in elongated open clusters; ray florets blue, pappus white.

Origin: Native.

Life form: Annual or biennial herb from a fibrous root.

Stems: Robust, erect, branched above, leafy; to 7 feet tall.

Leaves: Blades variable, elliptic to triangular in outline, irregularly deeply cut, grayish green above, veins often hairy below, margins toothed; to 12 inches long.

Flowers: Heads small, blue to occasionally white; involucral bracts vase-shaped, green with purple tips; ½ inch wide. June through September.

Fruits: Achenes brown, ribbed, beaked with white hairs.

Distribution: Moist open woods, woodland edges, thickets, roadsides, waterways. Common.

In Kentucky: AP, (IP-especially in outer Bluegrass), ME.

The height of this summer-flowering plant is impressive and usually attracts attention more than the small blue flowers.

The genus name, *Lactuca,* is from the Latin *lactus* "milk" and refers to the milky sap that exudes from the plant when cut. Native Americans would use the juice to cure the blisters from poison-ivy.

Purple-rocket

Mustard Family
Iodanthus pinnatifidus (Michx.) Steud.
Brassicaceae

Key features: Basal and lower leaves pinnately lobed, upper ones narrow, toothed; flowers with 4 purplish pink to white petals that are narrowed at base.

Origin: Native.

Life form: Perennial herb from a taproot.

Stems: Erect, simple to branching above, smooth, ridged; 1 to 3 feet tall.

Leaves: Stem leaves alternate, reduced upwards, blades lanceolate, oblong to elliptic, veins pale below; to 6 inches long.

Flowers: In terminal and axillary clusters from upper leaves; sepals 4, pale violet below, tips green; petals 4, spreading or reflexed; stamens 6, anthers purple; pistil 1, purple; 1/2 inch long. May through June.

Fruits: Pods ascending to spreading, linear, beaked, to 1 inch long; seeds tiny, in 1 row.

Distribution: Moist open woods, limestone ledges. Rare.

In Kentucky: IP, ME.

This beautiful late-spring wildflower is a welcome sight in shady woods. The genus name, *Iodanthus,* is from the Greek *iodes* "violet-colored" and *anthos* "flower." This species was rare in 1941 as recorded by Mabel Slack in her master's thesis on the flora of Cherokee Park and is still rare today.

Tall bellflower

Bellflower Family
Campanulastrum americanum (L.) Small
Campanulaceae
(*syn=Campanula americana* L.)

Key features: Leaves alternate, narrow, finely toothed; flowers star-shaped, blue-violet with white ring in center with protruding curved style, petal tips slightly twisted or curled.

Origin: Native.

Life form: Annual or biennial herb from a taproot.

Stems: Single, sometimes with a few side stems, slightly hairy; 2 to 5 feet tall.

Leaves: Blades ovate to lanceolate, dull green above, bases tapering into a slender stalk, tips long-tapering; 2 to 6 inches long.

Flowers: Solitary or clustered in upper leaf axils; sepals green, tubular with 5 narrow lobes; petals 5, ovate, spreading; stamens 5; pistil 1, style blue with tip curved upward; 1 inch wide. July through September.

Fruits: Capsule 5-angled.

Distribution: Wooded slopes, limestone ledges. Uncommon.

In Kentucky: AP, IP, ME.

This beautiful summer wildflower was used by Native Americans as a leaf tea for coughs and tuberculosis.

Indian tobacco

Bell flower Family
Lobelia inflata L.
Campanulaceae

Key features: Plant poisonous; stems with milky sap; leaves alternate; flowers tubular, light blue to white; fruits inflated.

Origin: Native.

Life form: Summer annual from a taproot.

Stems: Erect, angular, hairy, often branched above; to 1½ feet tall.

Leaves: Blades lanceolate to ovate, margins slightly toothed, lower ones short-stalked, upper sessile, becoming reduced to small leaflike bracts; 2 to 3 inches long.

Flowers: In terminal or axillary racemes; petals tubular with 5 spreading lobes, upper lip 2-lobed, lower lip 3-lobed; stamens 5, fused by anthers; ¼ to ½ inch long. July through October.

Fruits: Capsule round, enclosed by inflated green sepals; seeds many.

Distribution: Disturbed ground in fields, open woods, thickets, roadsides. Common.

In Kentucky: AP, IP, ME.

Also called asthma weed, this plant was used by the Native Americans for respiratory ailments, asthma, whooping cough, and chronic bronchitis.

Another popular name is pukeweed. The plant was used in folk medicine to induce vomiting and sweating. The species name, "inflata," refers to the base of the flowers, which inflate around the seed capsule.

Common dayflower

Spiderwort Family
Commelina communis L.
Commelinaceae

Key features: Stems weak; leaves alternate, entire; flowers with 2 large upper blue petals and 1 small, lower, white petal, in clusters opening 1 at a time.

Origin: Asia.

Life form: Summer annual from fibrous roots.

Stems: Slender, erect to ascending or creeping, rooting at the lower nodes; 1 to 3 feet tall.

Leaves: Leaf stalks with membranous sheath at base; blades ovate-lanceolate, veins parallel, tips long-pointed; to 4 inches long.

Flowers: Sepals 3; petals 3; stamens 6, yellow, unequal; pistil 1; leaflike bract below flower heart-shaped. May through October.

Fruits: Capsule 2- or 3-valved; seeds brownish red, rough, very tiny.

Distribution: Disturbed moist ground along roadsides, woodland edges, thickets, waterways. Uncommon.

In Kentucky: AP, IP, ME. Listed as a Lesser Threat by the Kentucky Exotic Pest Plant Council.

Similar species: Creeping dayflower (*Commelina diffusa* Burn. f.) is a weak summer annual from the Old World. The reclining or creeping foliage is similar to those of common dayflower, but the **small flowers consist of 3 dark blue petals.** Cher, Iroq. Disturbed moist ground, roadside ditches. Uncommon. In Kentucky: (AP-rare), (IP-rare), (ME-common). May through October.

The genus was named by Carl Linnaeus, the great Swedish botanist, for three Dutch brothers named Commelin. Two of the brothers, represented by the upper petals, were well-known botanists, and the third brother, the insignificant lower petal, did nothing for science.

In the southwestern United States, the Navajo Indians prepared a drink from this plant and gave it to aged men and women to increase their potency. The tribes believed so strongly in the efficacy of this infusion that they also gave it to their stud animals.

The beautiful flowers bloom only for one day—hence the name dayflower.

Virginia spiderwort

Spiderwort Family
Tradescantia virginiana L.
Commelinaceae

Key features: Stems jointed; leaves long, narrow; flowers blue-violet, sepals densely hairy, lacking glands; two leaflike bracts above wider and larger than leaves.

Origin: Native.

Life form: Perennial herb often growing in clumps from thick fleshy roots.

Stems: Erect, smooth, internodes sometimes hairy; to 1½ feet tall.

Leaves: Alternate, sessile, often bending downward; blades linear-lanceolate, upper ¾ inch wide, dark olive green, veins parallel, margins smooth, bases tapering, tip long-tapering; 5 to 12 inches long.

Flowers: Mostly terminal, in small clusters, stalked; sepals 3; petals 3, rounded; stamens 6, yellow, filaments blue-bearded; pistil 1; 1 inch wide. May through June.

Fruits: Capsule small, splitting open into 3 parts; seeds oval to oblong, brown, tiny.

Distribution: Moist to dry woods, woodland edges. Uncommon.

In Kentucky: AP, IP, ME.

Similar species: Wide-leaved spiderwort (*Tradescantia subaspera* Ker Gawl) is a perennial herb to 3 feet tall. The **zigzagged stem** has alternate leaves with the **upper blades to 2 inches wide.** The bluish purple flowers have **3 hairy sepals that are** usually **glandular** and 3 petals. Cher. Moist woods, woodland borders. Uncommon. June through September. In Kentucky: (AP-rare), IP, ME.

Virginia spiderwort is an old-fashioned native garden plant that was first taken to England in Colonial times. The beautiful blue violet flowers open in the morning and close in early afternoon, when the petals wither and turn into a liquid. In heavy morning dew, the pigment colors the dew drop blue before it dries up and disappears—hence the common name widow's-tears. It is also called early spiderwort because it flowers earlier than the wide-leaved spiderwort.

The genus name honors John Tradescant, a botanist and gardener of Charles I of England.

American groundnut

Legume Family
Apios americana Medik.
Fabaceae

Key features: A vine from rhizomes bearing fleshy tubers; leaves alternate, pinnately compound; flowers pealike, purplish brown to mauve.

Origin: Native.

Life form: Perennial herbaceous vine.

Stems: Trailing and twining over other vegetation, smooth; to 10 feet long.

Leaves: Leaflets 5 to 7, blades ovate, margins entire, surface below smooth to hairy, tips pointed; 2 to 3 inches long.

Flowers: Solitary or paired, in dense axillary clusters; sepals 5, green; petals 5, upper one broad, reflexed, lateral ones smaller, lower petal coiled; $1/2$ inch long. July through September.

Fruits: Pod narrow, straight to slightly curved, coiled after opening, 2 to 4 inches long; seeds several.

Distribution: Moist woods, thickets, especially common along the River Walk at Shawnee.

In Kentucky: AP, IP, ME.

Also known as Indian potato, hopniss, turkey pea, and potato bean, this native species was considered one of the most famous edible wild plants of pre-European North America. High in starch and protein, the crunchy potato-like tubers have been an important food source for Native Americans, early explorers, and colonists in eastern North America. It was noted that the Pilgrims of Plymouth in 1623 survived on groundnut when their corn supply was exhausted. In 1654, the town of South-ampton in the Connecticut River Valley passed a law that prohibited the Amer-indians from digging this species on "English-land."

In Ireland, during the potato famine of 1845, groundnut was introduced into Europe as a possible replacement. However, cultivation was abandoned when growing potatoes became possible again. The tubers can be used in soups, stews, or fried like potatoes.

Apios is Greek for "pear" and refers to the pear-like tubers on the rootstock.

American hogpeanut

Legume Family
Amphicarpaea bracteata (L.) Fernald
Fabaceae

Key features: A twining vine; leaves pinnately compound; flowers in upper leaf axils pealike, purplish pink to whitish, basal flowers inconspicuous.

Origin: Native.

Life form: Perennial vine from fibrous roots.

Stems: Slender, hairs white to brown, appressed; to 4 feet long.

Leaves: Leaflets 3, blades ovate, the terminal leaflet larger than other 2, bases broadly rounded, tips blunt to sharp pointed, hairs appressed; 1 to 2 inches long.

Flowers: Few, in nodding axillary clusters on long stalks; flowers of 2 kinds: those in upper axils pealike, upper petal larger than lower 3; lower basal ones inconspicuous, lacking petals and on slender creeping stems. July through September.

Fruits: Pods bean-like, flat, oblong, pointed at the tips, coiled after opening; seeds 3 in normal flowers and 1 in petal-less flowers.

Distribution: Moist open woods, thickets. Common in Iroquois and Shawnee; uncommon in Chickasaw; rare in Cherokee.

In Kentucky: AP, (IP-mostly inner Bluegrass), ME.

The genus name is from the Greek *amphi* and *carpos,* which means "of both kinds of fruits," and refers to the two kinds of fruits produced on the plant: those in the leaf axils above and those from the petal-less flowers below. It was a very important food plant among Native Americans, and the fruits were used in traditional medicine. An infusion of root tea was blown on a snake bite to help heal the wound. The underground, nutritious fruits were said to have a pleasant taste and were used by the Cherokee Indians to make bean bread or added to cornmeal and hot water.

Stems: Erect to ascending, slender, hairs densely spreading or hooked on upper stems; to 4 feet tall.

Leaves: Leaflets 3, terminal leaf the largest, to 4 inches long, blades ovate, dark green above, paler below, margins entire, veins pale, bases rounded, tips blunt to pointed; leaflike stipules triangular to ovate-lanceolate, tips long-tapering, persistent.

Flowers: Many, in loosely branched axillary and terminal elongated clusters; sepals small, lobes equal or slightly longer than tube; petals purplish pink. May through July.

Fruits: Segments 4 to 6 (9), margins wavy above, notched below, hairy-margined; seeds 1 per segment.

Distribution: Low moist woods, woodland edges, thickets, roadsides ditches, fields. Common.

In Kentucky: AP, IP, ME.

Similar species: Few flowered tick-trefoil [*Desmodium pauciflorum* (Nutt.) DC.] has stems that are erect, ascending or lying on the ground. The alternate leaves are 3-foliate with ovate blades. There are only a **few, white, pealike flowers** produced on a stalk 4 to 8 inches long. Fruit pods with **1 to 3 triangular, deeply lobed segments.** Iroq. Moist open woods, along trails, woodland edges. Uncommon. June through August. In Kentucky: AP, IP, (ME-rare).

Hoary tick-trefoil

Legume Family
Desmodium canescens (L.) DC.
Fabaceae

Key features: Stems hairy; leaves alternate, leaflets 3; flowers purplish pink, pealike; fruits in triangular segments of 4 to 6 (9).

Origin: Native.

Life form: Perennial herb from a rhizome.

Stems: Loosely tufted, bright green, flattened, about the same width as the leaves; 6 to 15 inches tall.

Leaves: Basal, blades long-linear, margins entire, parallel-veined, bluish green; 4 to 24 inches tall.

Flowers: Terminal, solitary or in few-flowered clusters; tepals 6, bristle-tipped; stamens 3; pistil 1; $1/2$ to 1 inch wide; leaflike spathes terminating stems, enclosing the flowers. May through July.

Fruits: Capsule round, 3-angled, purplish-tinged; seeds round, numerous.

Distribution: Open woods, fields. Uncommon.

In Kentucky: AP, IP, ME.

Narrowleaf blue-eyed grass

Iris Family
Sisyrinchium angustifolium Mill.
Iridaceae

Key features: Stems winged; leaves grasslike; flowers blue with darker blue lines, center yellow or green.

Origin: Native.

Life form: Perennial herb with fibrous roots.

The Cherokee Indians would steep the roots and give to children to help stop diarrhea, and the greens were cooked and taken to regulate the bowels. The Iroquois Indians made a decoction from the roots and stalks that was taken in the morning before meals to help constipation.

The flowers bloom in the morning sun and only for one day. Other common names are satin-flower, blue-eyed lily, and star-eyed grass.

American false pennyroyal
Mint Family
Hedeoma pulegioides (L.) Pers.
Lamiaceae

Key features: Plant with a strong mint scent; leaves opposite, heavily glandular-dotted; flowers small, pale blue to violet, in whorls in the leaf axils.

Origin: Native.

Life form: Summer annual herb from fibrous roots.

Stems: Erect, single or branched, light green, slender, densely short-hairy; 4 to 16 inches tall.

Leaves: Blades oblong-ovate, margins entire or slightly toothed, bases tapering, tips pointed; to 1 inch long.

Flowers: Sepals, 5, ribbed; petals tubular, weakly 2-lipped, upper lip flat, notched, lower lip 3-cleft, spreading; stamens 2; each pair of flowers subtended by a small leaflike bract. July through September.

Fruits: Nutlets 4.

Distribution: Dry woods, along trails, roadsides. Rare in Cherokee; common in Iroquois.

In Kentucky: AP, IP, ME.

This small but strongly aromatic plant is also called pennyrile and named after the Pennyroyal region in the western part of the state. The generic name is from the Greek *hedys* "sweet" and *osme,* "scent."

The Cherokee Indians used the beaten leaves as a poultice to place over a toothache. Leaves rubbed on the skin were said to keep insects from biting, especially ticks.

Perilla mint

Mint Family
Perilla frutescens (L.) Britton
Lamiaceae

Key features: Plant rank-smelling; leaves opposite, green to purplish bronze; flowers purple or white, small in loose, elongated clusters.

Origin: India.

Life form: Summer annual from fibrous roots.

Stems: Stout, purplish bronze to green, 4-angled; up to 3 feet tall.

Leaves: Long-stalked, blades oblong to broadly ovate, veins distinct, coarsely toothed, often crisped, tips pointed; to 5 inches long.

Flowers: In terminal and axillary clusters; sepals tubular, 2-lipped, densely white hairy; petals 5, rounded, lowest one largest; stamens 4, anthers purple; style divided; involucral bract below each flower. July through September.

Fruits: Nutlets 4, rounded, net-veined, enclosed by the enlarged, papery sepals.

Distribution: Disturbed ground in moist shady woods, fields, roadsides. Common.

In Kentucky: AP, IP, ME. Listed as a Significant Threat by the Kentucky Exotic Pest Plant Council.

For centuries, this plant was cultivated in China, but today it is mostly grown in Japan and Korea, where it is an important garnish and vegetable, often used like parsley.

The seeds are an important source of oil and are rich in linolenic acid. This oil is used in the manufacture of printer's ink, artificial leather, and water-proofing for clothing.

Several garden varieties have been found in the parks: those with a deep bronze color or with leaves either crisped, cut, or wrinkled. Another common name is beefsteak plant.

Self-heal

Mint Family

Prunella vulgaris L. var. *lanceolata*
Lamiaceae

Key features: Stems 4-angled; leaves opposite, blades lanceolate with leaf bases gradually narrowed or wedge-shaped; flowers purple, blue, or pink in dense cylindrical clusters.

Origin: Native.

Life form: Perennial herb from fibrous roots and stolons.

Stems: Either erect, ascending, or creeping, simple or branched; to 2 feet tall.

Leaves: Lower long-stalked; blades with distinct midvein below, margins entire to toothed, tips pointed to blunt; 1/2 to 4 inches long.

Flowers: In terminal and axillary clusters; sepals green or purple, tubular, nerved, tips long-pointed; petals 2-lipped, the upper lip arched, lower lip 3-lobed; involucral bracts rounded, membranous, bristly hairy, subtending each cluster of 3 flowers. May through October.

Fruits: Nutlets 4, obovate, brownish, slightly flattened, tapering to a white point.

Distribution: Disturbed ground in open woods, turf, fields, roadsides, thickets, waterways. Common.

In Kentucky: AP, IP, ME.

This variety is also called American heal-all and has lanceolate leaves that taper at the ends. The European variety *Prunella vulgaris* var. *vulgaris* has a broader, oval-shaped leaf with rounded leaf bases. Both varieties thrive in disturbed sites and can occur together. Herbarium specimens collected within the state are mostly of the native variety, and only a few records are of the European weed.

The name heal-all refers to the belief that the plants could be used to heal almost any ailment. The name *Prunella* is from the medieval name, *Brunella,* and refers to *Braune,* German for "quinsy," an uncommon but serious tonsil infection that the plant was said to cure.

Like most mints, the leaves are aromatic and the flowers yield abundant nectar, an important food source for bumblebees.

Chocolate vine
Lardizabala Family
Akebia quinata (Houtt.) Decne
Lardizabalaceae

Key features: A semi-evergreen woody vine or ground cover; leaves alternate, circular in outline, divided into 5 leaflets; flowers plum-colored.

Origin: China, Korea, Japan.

Life form: Perennial vine from rhizomes.

Stems: Climbing, trailing, green when young and turning brown when mature, ridges wavy, dotted with raised bumps; to 80 feet long.

Leaves: On slender stalks to 3 inches long; blades divided into 5 obovate to elliptic leaflets, midvein distinct, net-veined below, tip rounded, notched: ½ to 3 inches long.

Flowers: Male and female flowers in separate axillary clusters: female flowers plum colored to brown, 1 inch wide; male flowers small, lighter purplish plum, anthers 5, dark purple. May.

Fruits: Pods oblong, purplish, 2 to 4 inches long; seeds black, tiny, imbedded in fleshy pulp.

Distribution: Disturbed woods, woodland edges, locally invasive at Cherokee: gaining hold in Iroquois at 2 woodland locations.

In Kentucky: Collected in 3 counties: Jefferson, Bourbon, Rockcastle. Cherokee Park has the largest population to date. Listed as a Significant Threat by the Kentucky Exotic Pest Plant Council.

Also known as akebia or five-leaf akebia, this perennial vine is rare in the state except for areas noted above, where it is destructive. Although mentioned in Mabel Slack's 1941 master's thesis on the flora of Cherokee Park, the first herbarium specimen in Jefferson County dates from 1956 and was collected in Cherokee.

This ornamental plant was introduced into the United States in 1845. It spreads vegetatively and is said to be restricted only by the height of the object it is growing on. Some reports say that it can grow to eighty feet in a single growing season, thriving in many diverse habitats. Forming dense, tangled infestations, akebia will kill ground vegetation as well as canopy trees by overtopping and smothering them in a tent-like fashion.

Pickerel-weed
Pickerel-weed Family
Pontederia cordata L.
Pontederiaceae

Key features: Plant aquatic; leaves narrow to arrow-shaped or heart-shaped; flowers blue violet with 2 greenish yellow dots on uppermost lip.

Origin: Native.

Life form: Emergent aquatic plant from creeping rhizomes often forming colonies.

Stems: Stout, erect, succulent, light green; 1 to 3 feet tall.

Leaves: Basal, erect, on long hollow, rounded stalks; blades variable in shape, waxy, succulent, veins parallel, margins smooth; to 10 inches long.

Flowers: Many, in a dense terminal spike above 1 leaf; petals 6, funnel-shaped, 2-lipped, each 3-lobed; stamens 6; stigma 1 to 2; $1/2$ inch wide. June through September.

Fruits: Capsule oblong, bladder-like; seed 1.

Distribution: Pond margins and in shallow, still water. Rare with small, local populations.

In Kentucky: (IP-rare), (ME-rare). Threatened status.

This striking aquatic plant is considered a threatened species as documented by the Kentucky Natural Heritage Database. It was first found growing in a quarry near Cave Hill Cemetery in 1940 and later was collected in a swamp south of Louisville. Since then, it has never been seen again in Jefferson County and is now only found in three counties: Fulton, Henderson, and Greenup.

Like other emergent aquatic plants, after blooming, the upright flower stalk curves downward and the fruit matures below the surface of the water.

Pickerel-weed belongs in a small family of freshwater aquatics. It is related to water hyacinth (*Eichhornia crassipes*), which is considered to be one of the world's most serious aquatic weeds, and both are available in the nursery trade. Plants in both parks have been planted.

Broad-leaved bluets

Madder Family
Houstonia purpurea L.
Rubiaceae
[*syn=Houstonia lanceolata* (Poir.) Britton]

Key features: Stems 1 to many; basal leaves absent at flowering time, opposite above; flowers small, lavender to white, funnel-shaped.

Origin: Native.

Life form: Perennial herb from fibrous roots.

Stems: Erect, simple or slightly branched, ridged, hairs few, stiff, white; 6 to 15 inches tall.

Leaves: Blades ovate to lanceolate, veins 3 to 5, margins entire to slightly irregularly toothed, bases sessile or tapering into a very short stalk; ¹/₂ to 1¹/₂ inches long.

Flowers: Numerous, produced in clusters about 1 inch wide in upper leaf axils; sepals 4, green, narrow; petals tubular with 4 spreading lobes. May through July.

Fruits: Capsule round, sepals persistent, narrow.

Distribution: Damp to dry woods, mossy slopes, shady roadsides. Uncommon.

In Kentucky: AP, IP, ME.

This attractive wildflower is also called Venus' pride, summer bluet, and large houstonia.

Veined skullcap

Mint Family
Scutellaria nervosa Pursh.
Lamiaceae

Key features: Stems square; leaves opposite, clasping to short-stalked; flowers light blue, solitary, on densely hairy stalks in the leaf axils.

Origin: Native.

Life form: Perennial herb from rhizomes and stolons.

Stems: Slender, erect or ascending, simple to slightly forking below; 8 to 15 inches tall.

Leaves: Blades ovate, margins toothed, veins prominent below, bases wedge-shaped to rounded, tips rounded to tapering; 1/2 to 1 1/2 inches long; upper leaves gradually reduced, margins slightly toothed to entire.

Flowers: Sepals 5, short-lobed, light green, hairy; petals tubular, 2-lipped, lower lip purple-dotted; stamens 4. May through June.

Fruits: Nutlets yellowish tan.

Distribution: Moist woods, shaded roadsides, mossy banks. Common.

In Kentucky: (AP-rare), IP, (ME-rare).

The common name comes from the resemblance of the two-lipped flowers, with the uppermost lip arched and humped reminding early botanists of the skullcap worn by the Romans. *Scutellaria* comes from the Latin *scutella* and refers to the humped lip that was likened to a rounded platter.

Members of this genus lack the typical mint smell; they are bitter to the taste and were used by early herbalists to treat fevers.

Horse-nettle
Nightshade Family
Solanum carolinense L.
Solanaceae

Key features: Plant poisonous; stems and leaves with stout, straw-colored prickles; flowers bluish violet or white, star-shaped; fruits yellow-orange, round berries.

Origin: Native.

Life form: Perennial herb from thick, creeping rhizomes.

Stems: Erect, greenish purple, loosely branched, prickly, hairs star-shaped; 8 to 24 inches long.

Leaves: Alternate, stalked; blade elliptic-oblong to ovate, margins irregularly toothed, lobed or wavy,

prickles straw-colored on midrib, veins, and leaf stalk; 2 to 7 inches long.

Flowers: In axillary clusters on prickly stems; sepals 5, green, narrow; petals 5, united; stamens 5, yellow orange, fused; pistil 1; 1 inch wide. May through September.

Fruits: Berries green, turning yellow-orange and wrinkly when mature; seeds circular, tiny, glossy yellow, 40 to 170 in juicy pulp.

Distribution: Disturbed ground in fields, roadsides, thickets, cultivated beds. Common.

In Kentucky: AP, IP, ME.

Similar species: Black nightshade (*Solanum ptycanthemum* Dunal ex. DC.) is a native summer annual that grows to 3 feet tall. The alternate leaves are triangular to elliptic with entire, wavy, or toothed margins. Small, **whitish yellow, star-shaped flowers** less than $1/16$ inch wide are produced in few-flowered axillary clusters. The **berries are lustrous green, round, turning black when mature.** Cher, Sen, Iroq, Shaw, Chick. Disturbed ground in open woods, roadsides, fields, thickets, cultivated beds, waterways. Common. May through September. In Kentucky: AP, IP, ME. (*syn=Solanum americanum* Mill.; *Solanum nigrum* L.)

Both horse-nettle and black nightshade contain solanine, a toxic alkaloid. Children especially are attracted to the unripe fruits, which are highly poison-ous. Symptoms from this poisoning include paralysis, salivation, abdominal pain, as well as circulatory and respiratory depression. Boiling the plant is said to destroy the toxic properties.

The Native Americans inhaled the smoke from the dried plants of black nightshade to treat toothaches.

Climbing nightshade
Nightshade Family
Solanum dulcamara L.
Solanaceae

Key features: Plant poisonous; vine climbing or trailing; leaves alternate, stalked, lobed; flowers purple, rarely white, star-shaped; berries bright red.

Origin: Eurasia.

Life form: Perennial, semi-woody vine from rhizomes.

Stems: Slender, purplish green to brown, woody at base; up to 12 feet long.

Leaves: Blades triangular in outline, usually 3-lobed, terminal lobe largest, smaller basal lobes 2, ovate to heart-shaped, margins entire, tips pointed; 1 to 4 inches long.

Flowers: In 6-to-12-flowered axillary or terminal clusters; sepals 5, persistent; petals 5, recurving; stamens 5, yellow, anthers united forming a cone surrounding 1 pistil, style extended just beyond anthers. June through September.

Fruits: Berries pulpy; seeds tiny, dull light yellow, circular.

Distribution: Open disturbed woods, thickets, roadsides, fields, often climbing over other vegetation. Uncommon.

In Kentucky: AP, IP, ME. Listed as Moderate Threat by the Kentucky Exotic Pest Plant Council.

The whole plant is poisonous and contains steroids and toxic alkaloids such as the deadly glucoside, solanine. The berries look like tiny cherry tomatoes and are attractive to children. Eating stems, foliage, and berries causes vomiting, convulsions, weakened heart, and paralysis.

The genus name, *Solanum,* comes from the Latin *solamen* "comfort, solace" and refers to the plant's soothing narcotic properties.

Smooth aster

Aster Family
Symphyotrichum laeve (L.) A. Love &
D. Love
Asteraceae
(*syn=Aster laevis* L.)

Key features: Stems smooth, slightly black-dotted; leaves alternate, bases clasping; flowers bluish purple in loosely branched panicles.

Origin: Native.

Life form: Perennial herb from stout rhizomes.

Stems: Erect, green to brown, ridged; to 30 inches tall.

Leaves: Blades ovate to lanceolate, smooth, margins entire to slightly toothed, midvein silvery white below, tips pointed; reduced upward, narrow; to 7 inches long.

Flowers: Inner disk florets many, yellow; outer ray florets bluish purple, narrow; involucral bracts in several series, tips greenish purple; $\frac{1}{2}$ to 1 inch wide. August through October.

Fruits: Achene linear, brown, tufted hairs pale.

Distribution: Dry to moist woods. Uncommon.

In Kentucky: (AP-rare), IP.

Native Americans who suffered from rheumatism would take a bath that contained the stems and flowers of fall blooming asters to help soothe the pain.

The old generic name, *Aster*, is Greek for "star" and describes the shape of the flowers.

Common blue heart-leaved aster

Aster Family
Symphyotrichum cordifolium (L.) G.L.
Asteraceae
(*syn*=Aster cordifolius L.)

Key features: Leaves alternate, ovate to heart-shaped, leaf stalk often slightly winged; flower heads bluish lavender and involucral bracts green with purple tips.

Origin: Native.

Life form: Perennial herb from rhizomes.

Stems: Erect, slender, hairs in lines; 1 to 3 feet tall

Leaves: Leaf stalk slender to slightly winged; blades rough above, margins sharply toothed, slightly to densely hairy below, tips pointed; upper stem leaves heart-shaped, reduced; 1 to 5 inches long.

Flowers: Heads many, in a loose, much-branched panicle; inner disk florets yellow to purplish; outer ray florets lavender to blue; involucral bracts narrow, overlapping, $^1/_2$ to $^3/_4$ inch wide. August through October.

Fruits: Achenes pale, smooth, 3 to 5 ridged.

Distribution: Moist woods, woodland edges. Common.

In Kentucky: AP, IP, ME.

Native Americans would use the root of the common blue heart-leaved aster to make an incense or smoke that had a peculiar scent that would attract deer so that they could then be shot with a bow and arrow at close range.

New England aster

Aster Family
Symphyotrichum novae-angliae
(L.) G.L. Nesom
Asteraceae
(*syn=Aster novae-angliae* L.)

Key features: Stems leafy at top; leaves alternate with clasping leaf bases; flower heads showy, deep purple to pinkish purple with yellow centers.

Origin: Native.

Life form: Perennial herb from creeping rhizomes and fibrous roots.

Stems: Stout, branching above, hairy, some glandular-tipped; 3 to 5 feet tall.

Leaves: Crowed, lower soon falling off; blades lanceolate, margins entire, tips pointed; 2 to 4 inches long.

Flowers: Heads several to many; inner disk florets small, yellow, tubular; outer ray florets purple; involucral bracts with purplish tips, loose, spreading, glandular-hairy; 1 to 2 inches wide. August through October.

Fruits: Achenes densely hairy, nerves obscure.

Distribution: Moist open woods, damp thickets at Cherokee; Summit Field at Iroquois. Uncommon.

In Kentucky: AP, (IP-mostly northern Bluegrass), ME.

This is one of a few plants that have a vibrant, rich blue color in the fall. Because of its beauty, this species is widely cultivated and sold in plant nurseries. It makes a great addition to any garden, be it formal or naturalistic.

Great blue lobelia

Bellflower Family
Lobelia siphilitica L.
Campanulaceae

Key features: Plant with milky sap; leaves alternate, narrow, sessile; flowers deep blue, occasionally white, 2-lipped, in elongated terminal clusters.

Origin: Native.

Life form: Perennial herb from a taproot and basal offshoots.

Stems: Erect, angular, smooth to sparsely hairy; 1 to 3 feet tall.

Leaves: Blades lanceolate to elliptical, midstem ones mostly lanceolate, margins toothed, sparsely hairy below, bases wedge-shaped, tips short-pointed; 2 to 5 inches long.

Flowers: Sepals 5, green, narrow, with earlike basal lobes; petals tubular, 2-lipped: upper lobes 2, erect, lower lobes 3, flaring outward; stamens 5 surrounding a pistil that projects upward through slit in upper lip; 1 inch long. August through October.

Fruits: Capsule round; seeds brown, shiny, net-veined, warty.

Distribution: Moist to wet ground along waterways, roadside ditches. Rare.

In Kentucky: AP, IP, ME.

This beautiful native wildflower is a welcome sight throughout Kentucky in late summer and fall. It is also a popular plant sold in the nursery trade because the eye-catching deep blue flowers add color to the cultivated landscape.

In Native American medicine, the leaves mixed with tobacco were smoked and used as a treatment for asthma, tonsillitis, and bronchitis. The roots were used to cure syphilis—hence the species name *siphilitica*. They were also used with food as a love medicine.

Stone mint

Mint Family
Cunila origanoides (L.) Britton
Lamiaceae

Key features: Stems smooth, 4-angled; leaves opposite; flowers purplish pink to white, weakly 2-lipped with 2 straight stamens protruding.

Origin: Native.

Life form: Perennial herb from slender rhizomes.

Stems: Brown, freely branched with multiple stems from a semi-woody base; to 16 inches tall.

Leaves: Sessile to very short-stalked; blades ovate to triangular, black-dotted, margins slightly toothed, bases rounded, tips pointed; ¾ to 1½ inches long.

Flowers: In short-stalked, axillary and terminal clusters; sepals 5-lobed, strongly 10-nerved, sparkly glandular-dotted; petals purple dotted inside, black to sparkly dotted outside; style purple, stigma 2-lobed; ½ inch long. August through October.

Fruits: Nutlets 4, smooth.

Distribution: Dry open woods in noncalcareous soils. Common.

In Kentucky: AP, IP, ME.

This fragrant member of the mint family was used in traditional medicine to treat headaches and fevers. It also was said to repel horseflies, and bunches were placed in the horse's bridle to keep them at a distance.

Green-dragon

Arum Family

Arisaema dracontium (L.) Schott.
Araceae

Key features: Plant from a corm; leaf solitary horseshoe-shaped held horizontally above the ground; flowers with a long tonguelike spadix.

Origin: Native.

Life form: Perennial herb.

Stems: Lacking except for the flowering stalk.

Leaves: Solitary, divided into 7 to 13 segments, central ones longest, outward progressively shorter, margins entire, narrowed at the base, tips pointed; on a long stalk to 3 feet tall.

Flowers: Spathe green, slender, slightly curved inward, 1 to 2 inches long; spadix long, projects beyond spathe, greenish white, with tiny male and female flowers hidden at base. April through June.

Fruits: Berries many, produced in large, rounded heads, green at first, turning bright reddish orange in late summer; seeds 1 to 3 per berry.

Distribution: Damp, shady woods, to open rocky woods. Rare.

In Kentucky: (AP-rare), IP, ME.

This unusual-looking plant is also known as dragon root and dragon-tail. It was used by early settlers to treat asthma. Native Americans used the root in making sacred bundles that gave the owner the power of supernatural dreams. Like Jack-in-the-pulpit, this plant contains calcium oxalate crystals, which cause a severe burning sensation and swelling of the throat when eaten.

leaflets unequal, tips sharp-pointed; 3 to 7 inches long.

Flowers: Small, borne on the club-like spadix, male flowers below, female flowers above or occasionally either all male or all female; hooded spathe, variable from light green to green with white or purple stripes, flap arching over the spadix. April through May.

Fruits: Berries green, turning brilliant glossy red by late summer, in rounded clusters; seeds 1 to 5, white to tan, in each berry.

Distribution: Moist woods. Common at Cherokee; rare at Seneca and Iroquois.

In Kentucky: AP, IP, ME.

Jack-in-the-pulpit

Arum Family
Arisaema triphyllum (L.) Schott.
Araceae

Key features: Plant from a deep bulb-like corm; leaves compound, 1 to 2, long-stalked; flowers with a club-like spadix and hooded spathe.

Origin: Native.

Life form: Perennial herb.

Stems: Erect, 1 to 3 feet tall and taller after flowering.

Leaves: Leaflets 3, terminal leaflet is larger than laterals, blades ovate to broadly rhombic, distinctly veined, margins entire, leaf bases of lateral

This spring wildflower is also known as Indian turnip and pepper turnip, describing the deep bulbous corm that has a peppery taste and is eaten like a potato. If not boiled, it causes an intense burning sensation and swelling of the lips, tongue, and throat due to calcium oxalate crystals. It was also named memory-root because naughty boys would play tricks by teasing each other to bite into the corm: the result being very painful and one they would never forget.

The common name, Jack-in-the-pulpit, refers to the hooded "pulpit" or spathe that wraps around and covers "Jack," the club-like "spadix." The plant smells of rotting meat and is pollinated by flies.

Common blue cohosh

Barberry Family
Caulophyllum thalictroides (L.) Michx.
Berberidaceae

Key features: Plant poisonous; stems bluish green; leaves ternately compound; flowers small, greenish yellow or greenish purple; berries dark blue.

Origin: Native.

Life form: Perennial herb with knotty rhizomes.

Stems: Erect, smooth; to 2½ feet tall

Leaves: Leaflets 3, obovate, 2-to-5-lobed, grayish green above, bluish gray below, stalked; 1 to 3 inches long.

Flowers: One-to-3-flowered, in terminal clusters; outer sepal-like bracts 3 or 4, tiny; sepals 6, petal-like, oblong to elliptic; petals 6, rudimentary, fan-shaped, hooded, brown, shorter than the sepals; stamens 6; pistil 1; about ¼ inch wide. April through May.

Fruits: Berry-like seeds 2, round, on thick stalks.

Distribution: Rich moist woods. Uncommon.

In Kentucky: AP, IP.

Also called papoose root or squaw root, this wildflower was used by the Native Americans and early settlers, who made a tea from the roots to facilitate childbirth. The Cherokee Indians made syrup from the roots that were given for "fits and hysterics."

The small flowers appear before the single leaf with its many leaflets fully open and ripen into attractive bright blue berry-like seeds. As the seeds enlarge, they burst through the ovary wall and develop in a completely exposed state.

Key features: Stems with a papery cylindrical sheath at each swollen joint; leaves with wavy to curled margins; flowers reddish brown at maturity; fruits with wings entire to slightly toothed.

Origin: Eurasia.

Life form: Perennial herb from a fleshy taproot.

Stems: Stout, ridged; 1 to 3 feet tall.

Leaves: Mostly basal, long-stalked; blades oblong-lanceolate, dull bluish green above, midvein distinct, bases round to heart-shaped, tips pointed; alternate stem leaves similar but smaller, bases tapering; to 15 inches long.

Flowers: In long, dense clusters, intermingled with narrow leaves; sepals 6, outer 3 small, inconspicuous, inner large, heart-shaped, bearing a plump grain-like structure at the base; all on short, dropping, threadlike stalks. April through July.

Fruits: Achenes reddish brown, triangular, margins of enlarged inner sepals (valves) entire or slightly toothed, veined; seeds small.

Distribution: Disturbed ground, especially in fields, roadsides, waterways, thickets. Common weed.

In Kentucky: AP, IP, ME.

Similar species: Broadleaf dock (*Rumex obtusifolius* L.) is a European perennial herb to 4 feet tall. It has **large, wide lower leaves with slightly wavy margins** and heart-shaped leaf bases; the alternate stem leaves are shorter and

Curly dock
Smartweed Family
Rumex crispus L.
Polygonaceae

narrower. Fruits are greenish purple with **4 to 8 bristly teeth on the margins of the enlarged, winglike, veined sepals.** Cher, Sen, Iroq, Shaw, Chick. Disturbed ground. Common weed. April through July. In Kentucky: AP, IP, ME.

The species within this genus are difficult to distinguish, and the fruits are essential for positive identification. All produce copious amounts of pollen and can cause hay fever.

The Iroquois Indians made a compound decoction that was given to induce pregnancy and as a tonic for women's health. Both the Cherokee and Iroquois Indians believed that the roots were good for purifying the blood. They also made a wash by putting the roots in vinegar that was used to treat ringworm.

The young leaves were boiled and eaten when food was scarce, and the seeds were roasted and drunk as a coffee substitute.

Early meadow-rue

Buttercup Family
Thalictrum dioicum L.
Ranunculaceae

Key features: Leaves alternate, 2 to 3 times divided; flowers greenish purple, in loose terminal and axillary clusters, either all male or all female.

Origin: Native.

Life form: Perennial herb from fibrous roots.

Stems: Erect, slender, smooth, greenish purple; to 2½ feet tall.

Leaves: Middle and upper long-stalked; leaflets 3 to 5, broadly ovate to kidney-shaped, dark green above, grayish below, margins bluntly lobed or toothed; up to ½ inch wide.

Flowers: Male with 4 sepals, oval to oblong, greenish purple; petals absent, stamens many, anthers yellowish green, drooping; female with 4 sepals, green or purple, small, firmer, petals absent, pistils several, purplish green, upright. April through May.

Fruits: Achenes ellipsoid, small, sessile, strongly ribbed.

Distribution: Rich moist woods, along waterways, limestone ledges. Rare.

In Kentucky: AP, IP, (ME-rare).

The genus, *Thalictrum,* is Greek for a plant with divided leaves, while *dioicum* means "two houses" and refers to the fact that male and female flowers grow on different plants.

The common name describes the early blooming period in spring.

Although some species contain thalictrine, a cardiac poison, the leaves are said to restore hair loss and prevent hair from falling out.

Green violet

Violet Family
Hybanthus concolor (T.F. Forst) Spreng.
Violaceae

Key features: Stems leafy; leaves alternate, tapering at base into a short stalk; flowers small, greenish white, nodding at the base of the leaf stalk.

Origin: Native.

Life form: Perennial herb from fibrous roots.

Stems: One to 2, erect, greenish brown, densely silky-hairy above; 1 to 3 feet tall.

Leaves: Blades elliptic, dark green above, veins pale, margins smooth to coarsely toothed, bases wedge-shaped, tips sharp,

long-tapered; 2 to 6 inches long; stipules green, narrow, to ¼ inch long.

Flowers: In the upper leaf axils, 1 to 3; sepals 5, green, linear with long, tapering tips; petals 5, oblong; stamens 5, united; pistil 1, club-like; April through May.

Fruits: Capsules widest in middle; seeds tan or mottled, tiny.

Distribution: Limestone cliffs in thin soil. Rare.

In Kentucky: AP, IP, ME.

This unusual-looking member of the Violet Family is the only representative of the genus in the United States. The other species, about 150, are found in the tropics and subtropics.

The Iroquois Indians made an infusion from the roots and stems and fed it to mares with injured fetuses.

Hybanthus means "humpback flower" and refers to the drooping flowers, and *concolor* means "of one color," which describes the green sepals and petals.

Smooth pigweed

Amaranth Family
Amaranthus hybridus L.
Amaranthaceae

Key features: Leaves alternate, oval; flower parts indistinguishable, greenish yellow, in slender bristly fingerlike clusters; outer stiff bracts 1 to 3, equal to or up to twice as long as the 5 pointed sepals.

Origin: Eastern North America, Central and South America.

Life form: Summer annual from a fibrous taproot.

Stems: Erect, stout, ridged or lined in red or white, often reddish at base; 3 to 8 feet tall.

Leaves: Often drooping on stalks as long as or shorter than blades; blades dull green above, distinctly veined with glandular hairs below, margins entire to slightly wavy; 1 to 7 inches long.

Flowers: Either all male or all female in slender, crowded terminal fingerlike clusters up to 2 feet long and in short axillary clusters; sepals surrounding either 1 pistil or 5 stamens. July through October.

Fruits: Urticle papery; seeds tiny, circular, reddish black, shiny.

Distribution: Disturbed ground in moist fields, thickets, roadsides, waterways. Uncommon.

In Kentucky: AP, IP, ME.

Similar species: Rough pigweed (*Amaranthus retroflexus* L.), also known as **redroot pigweed,** is a summer annual with stout, erect, light green stems that are often branched above. The alternate, long-stalked leaves are oval, dull green above and have wavy margins. The flowers, either all male or all female, are produced in several **broad, crowded, stout spikes. The outer stiff bracts are 2 to 3 times as long as the rounded or squarish sepals.** Cher, Iroq, Shaw. Moist fields, roadsides, thickets. Uncommon. July through October. In Kentucky: AP, IP.

Several species closely related to smooth pigweed were grown as a major food crop of the Aztecs in Mexico. Even today, amaranths are grown and harvested in parts of Mexico and the southwestern United States. Cooked like spinach or eaten as a fresh vegetable, the leaves are rich in protein, vitamins, and iron. It can also be ground into flour.

Some archeologists suggest that the Ozark-Bluff dwellers, who lived in northern Arkansas and Missouri, may have domesticated this species, as the seeds have been found in archaeological remains.

Rough pigweed is more common west of the Mississippi River, where it is a major weed in cultivated crops. Smooth pigweed is more common east of the Mississippi River. Both are known to hybridize, thus making identification difficult.

Winged sumac
Sumac Family
Rhus copallina L.
Anacardiaceae

Key features: Shrub with alternate, compound leaves, leaflets narrow, margins entire, attached to a central winged stalk; berries deep crimson red, in drooping clusters.

Origin: Native

Life form: Woody compact shrub when young, becoming more treelike and open as matures; to 16 feet tall.

Leaves: Leaflets 7 to 29, shiny dark green, lanceolate, 1 to 3 inches long, bases slightly unequal, hairy below.

Flowers: Small, in dense terminal clusters 4 to 8 inches long; sepals 5-parted; petals 5, greenish white to yellow, spreading; stamens 5. June through August.

Fruits: Berries small, 1-seeded, hairy, in drooping terminal clusters that turn black and last into winter.

Distribution: Fields, thickets, open disturbed woods, woodland edges. A dominant shrub at Summit Field in Iroquois, forming extensive colonies; uncommon in other parks.

In Kentucky: AP, IP, ME.

Similar species: Smooth sumac (*Rhus glabra* L.) is a native shrub to 15 feet tall with a wide, spreading crown. It has **smooth, tannish purple, rounded twigs** with 7 to 31 narrow **toothed leaflets** that are **smooth**, distinctly veined, long-pointed at the tips, **and lacking a winged stalk.** The small greenish white flowers are produced in dense clusters and are either all male or all female. In fall, the **erect fruit clusters turn a brilliant scarlet red** and are covered with sticky, appressed hairs. Cher, Sen. Roadsides, woodland edges. Uncommon. May through June. In Kentucky: AP, IP, ME.

The genus *Rhus* contains many members that are grown ornamentally for their brilliant reddish orange fall foliage. Because they tolerate dry, sunny, disturbed sites, they are often planted along highways as well as in the home landscape.

Winged sumac is less aggressive than smooth sumac, but both can spread rapidly by suckering. The ecological value to wildlife is high, as they provide habitat and food to songbirds, rabbits, deer, and fox, especially in winter. Both prefer areas that have a history of disturbance and can become quickly established after a fire.

The Cherokee Indians used the red and black berries for a dye. They also were used to aid in bed-wetting. The Native Indians in the northeast used the changing leaf color in the fall as an indication that the sockeye salmon were spawning.

Poison-ivy

Sumac Family
Toxicodendron radicans (L.) Kuntze
Anacardiaceae
(*syn=Rhus radicans* L.)

Key features: Plant woody, erect or viny; leaves alternate, long-petioled with 3-leaflets, the middle leaflet long-stalked; flowers greenish yellow, in slender axillary clusters.

Origin: Native.

Life form: Perennial vine or shrub from rhizomes.

Stems: Warty, when climbing they are supported by aerial roots.

Leaves: Blades ovate to elliptic, shiny, smooth above, short-hairy on the midrib below, margins entire, toothed or lobed, bases rounded or wedge-shaped, tip tapered to a point; to 5 inches long.

Flowers: Male and female flowers separate; sepals 5, green with yellow margins; petals 5, greenish yellow, ovate to oval; stamens 5, anthers orange in male flowers; ovary rounded, stigmas yellowish white in female flowers. May through July.

Fruits: Grayish white to tan, ¼ inch or less, each with a striped, 1-seeded stone.

Distribution: Disturbed ground, especially in open woods, fields, thickets. Common to abundant.

In Kentucky: AP, IP, ME.

The genus name, *Toxicodendron,* means "poison tree." The plant contains an oily resinous compound, urushiol, that can cause inflammation of the skin, blistering, and itching in people who are sensitive to poison-ivy. Others can handle it with little or no reactions. It is said that urushiols can remain active on objects such as tools, pets, and clothing—and even in dead plants—for over a year.

A Native American remedy consists of squeezing the sticky juice of jewelweed or touch-me-not on the affected areas until they heal. This practice is still used today.

Cluster sanicle

Carrot Family
Sanicula odorata (Raf.) Pryer & Phillipe
Apiaceae
(*syn=Sanicula gregaria* E.P. Bicknell)

Key features: Leaves palmately divided, toothed; flowers tiny, in rounded umbels, greenish yellow with long styles; fruits with hooked bristles.

Origin: Native.

Life form: Perennial herb with fibrous roots.

Stems: Smooth, purplish green, ridged, occasionally branched; to 2 feet tall.

Leaves: Basal and lower ones long-stalked; blades palmately divided into 5 leaflets, margins cleft to coarsely toothed, bases wedge-shaped, tips pointed, 2 to 4 inches long; upper leaves divided into 3 leaflets, similar, smaller, short-stalked to sessile.

Flowers: Umbels with tiny stalked flowers, a few perfect, the rest male; sepals 5, green; petals 5; stamens 5, exserted; anthers bright yellow; ovary bristly, styles longer, recurved. June through August.

Fruits: Rounded, small, bur-like, styles green, slender, longer than bristles.

Distribution: Moist to dry woods, shady roadsides, limestone ledges, waterways. Common.

In Kentucky: AP, IP, ME.

This common wildflower with nonshowy flowers has bitter foliage, and white-tailed deer and other mammals avoid eating this plant because of the taste.

Giant ragweed

Aster Family
Ambrosia trifida L.
Asteraceae

Key features: Plant tall; leaves mostly opposite, unlobed to deeply 3-to-5-lobed; flower heads greenish yellow, mostly in terminal clusters.

Origin: Native.

Life form: Summer annual herb from a fibrous taproot.

Stems: Erect, reddish green, rough-hairy, branching above; to 13 feet tall.

Leaves: Stalked, blades ovate to elliptic, surfaces rough, margins toothed; to 12 inches long.

Flowers: Separate male and female flowers on same plant, all greenish yellow: female flowers in the leaf axils below; male flowers showier, borne in terminal racemes. July through October.

Fruits: Achenes obovoid, black, tiny, central beak surrounded by 4 to 8 projections.

Distribution: Disturbed ground, especially in moist areas along Beargrass Creek in Cherokee and the River Walk at Shawnee. Abundant.

In Kentucky: AP, IP, ME.

Similar species: Common ragweed (*Ambrosia artemisiifolia* L.) is a native summer annual herb that **grows to 3 feet tall.** The leaves are **alternate above, opposite below and deeply divided into many bluntly toothed or lobed segments.** The female flowers are in the leaf axils below, and the male flowers are borne in small inverted clusters in terminal racemes, all on same plant. Cher, Sen, Iroq, Shaw, Chick. Disturbed ground. Abundant. July through October. In Kentucky: AP, IP, ME.

Carried by the wind, pollen from these two species in late summer and early fall is an enemy of hay fever sufferers. Long before allergenic preparations were available, common ragweed was used as an astringent to open closed nasal passages and to relieve constant sneezing.

The crushed leaves were used by the Cherokee Indians to cure the sting from insect bites as well as hives. The Iroquois Indians made an infusion from the roots that was taken for strokes. Both species are considered troublesome weeds throughout Kentucky.

The genus name, *ambrosia,* means "the food of the Gods." Giant ragweed seeds have been found in prehistoric archeological sites. The seeds were cultivated by pre-Columbian Indians and found to be four to five times larger than those today.

Mugwort

Aster Family
Artemisia vulgaris L.
Asteraceae

Key features: Plant clump-forming; leaves aromatic, pinnately lobed, white-woolly below; flowers inconspicuous, in leafy terminal spikelike clusters.

Origin: Eurasia.

Life form: Perennial herb from stout rhizomes.

Stems: Erect, often reddish brown, ridged, angular, becoming woody with age; to 5 feet tall.

Leaves: Alternate, 2 to 4 inches long; blades with unequally cleft segments nearly to midrib, dark green above,

margins entire to toothed, veins distinct; leaflike appendages, 1 to 2 pairs, at base of leaf stalk.

Flowers: Heads many, small; inner disk florets greenish yellow, turning purplish green when mature; involucral bracts narrow, cottony. July through October.

Fruits: Achenes brown, ridged, oblong, minute bristles at tip; seeds slightly curved, brown to yellowish brown with silvery, shiny stripes.

Distribution: Disturbed open ground, especially invasive along River Walk at Shawnee: spreading along Beargrass Creek in open turf at Cherokee. Uncommon elsewhere.

In Kentucky: AP, IP, ME. Listed on Watch List by Kentucky Exotic Pest Plant Council.

Brought over to North America by the European settlers in the mid-1800s, this plant with its aromatic foliage is often confused with that of common ragweed. This clump-forming, aggressive plant is a major weed found growing in nursery fields, orchards, vineyards, and gardens. The persistent rhizomes break off easily and regenerate quickly.

It has spread by way of contaminated topsoil, farm equipment, and burlap-wrapped nursery stock that is infested with rhizomes. The seeds spread by wind and water and are known to germinate after hundreds of years in the soil.

Invasive plant
Cher, Sen, Iroq, Shaw, Chick

In Ireland, smoked mugwort wreaths were unearthed from Irish archeological sites. These wreaths were hung over doors and believed to provide protection from evil spirits and wild animals.

Other common names are green ginger, mugweed, and sailor's tobacco. Mugwort refers to the plant's use as an insect repellent against gnat-like midges.

Asian bittersweet

Bittersweet Family
Celastrus orbiculatus Thunb.
Celastraceae

Key features: Woody climbing vine; leaves alternate, roundish, glossy green above; flowers greenish yellow; fruit capsules orange-yellow, dangling in axillary clusters.

Origin: Northeastern Asia.

Life form: Deciduous vine from shallow bright orange roots.

Stems: To 60 feet high, grayish white when young, turning dark brown and ridged when mature.

Leaves: Simple, stalked; blades round to oblong-ovate, margins wavy to bluntly toothed, tips blunt to short-pointed; 2 to 5 inches long.

Flowers: Male and female on separate plants in axillary clusters: male with 5 small outer sepals; petals 5; stamens 5, about equal in length to petals, pistil minute; female flowers similar but with 5 minute stamens and a 3-lobed stigma. May through June.

Fruits: Capsules round, 3-valved, splitting open to expose 3 bright red-orange seeds in fall.

Distribution: Disturbed ground in open or shady woods, woodland edges, thickets, fields. Abundant, but being eradicated due to its invasive tendencies.

In Kentucky: (AP-rare), IP, (ME-rare). Listed as a Severe Threat by the Kentucky Exotic Pest Plant Council.

Also known as Asiatic bittersweet, Oriental staff vine, climbing spindle berry, and round-leaved bittersweet, this vine was introduced into the United States in the late 1800s as a garden ornamental with showy reddish orange berries that ripen in autumn. Often planted around old homesites, it has escaped and spread, climbing up and over other vegetation forming dense, tangled, heavy masses that are capable of girdling stems and uprooting trees. It produces an abundance of seed that is dispersed by birds and small mammals.

Asian bittersweet is displacing our native species, American bittersweet, through competition for habitat and hybridization.

American bittersweet

Bittersweet Family
Celastrus scandens L.
Celastraceae

Key features: Climbing woody vine; leaves alternate, ovate; flowers greenish yellow; fruit capsules orange-yellow, dangling in terminal clusters.

Origin: Native.

Life form: Deciduous vine from a taproot and root suckers.

Stems: To 30 feet or more, light brown, tendrils absent.

Leaves: Simple, slender stalked; blades ovate to elliptical, dark green above, margins finely toothed, bases rounded or tapered, tips pointed; 2 to 4 inches long.

Flowers: Male and female flowers on separate plants, similar to Asian bittersweet. May through June.

Fruits: Capsule round, 3-valved, splitting open in fall to reveal the scarlet to crimson fleshy seeds.

Distribution: Woodland edges along roadsides, often climbing up trees. Rare.

In Kentucky: AP, IP.

The name bittersweet was given to this attractive native vine by the European colonists in the eighteenth century because the showy fruits resembled those of the weedy bittersweet (*Solanum dulcamara*) in the Nightshade Family; both are poisonous to humans.

The bright red berries are highly prized for making wreaths and other indoor decorations. Over-collecting of the berry-laden branches has reduced wild populations in some areas significantly. In addition, the nonnative Asian bittersweet has escaped cultivation and is attaining a similar geographical range. Identification on leaf character alone is not reliable. It is important to look at the flower inflorescence: native bittersweet has terminal clusters and the nonnative species has axillary clusters.

Hearts-a-bursting-with-love

Bittersweet Family
Euonymus americanus L.
Celastraceae

Key features: Understory shrub, twigs 4-angled; leaves opposite, waxy green; fruits bright reddish pink, 3-to-5-lobed, warty.

Origin: Native.

Life form: Upright or sprawling with stiff twigs; to 7 feet tall, often suckering.

Leaves: Simple, short-stalked; blades narrow to ovate, bright green above, pale below, smooth, margins finely toothed, tips pointed; 1 to 3 inches long.

Flowers: Solitary or in axillary clusters of 2 to 3 on slender stalks; sepals 5, green; petals 5, greenish purple, rounded, spreading, flat; stamens short; stigma 3 to 5 lobed; about ⅓ inch wide. May through June.

Fruits: Capsules splitting open to reveal 4 shiny orange-red berry-like seeds in late August though September.

Distribution: Moist woods, woodland edges, ravine slopes, waterways. Common at Iroquois; rare at Cherokee, Shawnee and Chickasaw.

In Kentucky: AP, (IP-but absent from northern part), ME.

Also called bursting heart, strawberry bush, and cat's paw, the unique fruits "burst" open in fall and are a colorful addition to the woods.

When young, the dark green, waxy plant trails over the ground and spreads rapidly. However, when it ages, the plant becomes upright with thicker stems and branches. These young plants can be seen at Iroquois and are often confusing to identify at first, looking more like a ground cover.

The Cherokee Indians made several infusions from the roots and bark that were used for ailments such as a stomach ache, breast complaints, sinus problems, and irregular urination.

This beautiful native shrub was once very common but is less so now due to browsing by overpopulated deer and infrequent reproduction.

Climbing euonymus

Bittersweet Family

Euonymus fortunei (Turcz.) Hand.-Mazz

Celastraceae

Key features: Evergreen groundcover, mat-forming or high climbing vine; leaves opposite and subopposite, simple, waxy; flowers greenish white, small; fruit capsules pinkish red.

Origin: China.

Life form: Evergreen plant from rhizomes.

Stems: Densely covered with minute warts, climbing to 20 plus feet by means of aerial roots.

Leaves: Short-stalked, blades elliptic to oval, upper surface dark green, veins whitish, margins bluntly to sharply toothed; about 2 to 3½ inches long.

Flowers: In axillary clusters borne on slender stalks; sepals 4, light green; petals 4, rounded; stamens 4, white; style 1, small. June through August.

Fruits: Capsules smooth, 3-to-5-lobed, opening to expose the orange-covered seeds.

Distribution: Disturbed woods, around abandoned home sites, rocky places, turf. Locally invasive.

In Kentucky: (AP-rare), (IP-mostly Bluegrass Region). Listed as a Severe Threat by the Kentucky Exotic Pest Plant Council.

Also called wintercreeper, this aggressive plant was introduced into North America primarily as an ornamental groundcover in 1907. Forming dense mats in shady woods, it climbs on and over anything in its path by attaching to objects with numerous small roots arising from the warty stems. In autumn, the showy fruits are highly visible on climbing vines, whereas the ground-cover populations rarely flower or fruit. This species is spread by humans and birds.

Lamb's-quarters

Goosefoot Family
Chenopodium album L.
Chenopodiaceae

Key features: Stems tan-and-red-ridged; leaves alternate, with white mealy powder below, especially when young; flowers tiny, inconspicuous, greenish purple.

Origin: Eurasia.

Life form: Summer annual herb from a fibrous taproot.

Stems: Erect, branched above; 2 to 4 feet tall.

Leaves: Variable, lower ones long-stalked; blades broadly ovate to rhombic, margins wavy, toothed or entire; upper ones smaller, narrow, clasping, margins mostly entire; 2 to 4 inches long.

Flowers: In dense terminal and axillary clusters; sepals 5; petals absent; stamens 6; pistil 1; covered with a white mealy powder. July through October.

Fruits: Urticle containing 1 tiny, shiny black or reddish brown, lens-shaped seed.

Distribution: Disturbed ground in fields, thickets, open woods, roadsides. Common weed.

In Kentucky: AP, IP, ME. Listed as a Moderate Threat by the Kentucky Exotic Pest Plant Council.

Lamb's-quarters is considered one of the world's most troublesome weeds. It is a prolific seed producer, and one average-sized plant can produce over 70,000 seeds. The common name is thought to be a corruption of "lammas quarter," an ancient festival in Britain.

This plant has been used since prehistoric times. In Alberta, Canada, a Blackfoot archaeological site was found with a storage room containing four to five liters of cleaned lamb's-quarters seeds. Today, the tender young plants are collected by the Native Americans and others in the southwestern United States, who boil and eat them like spinach; the leaves are rich in vitamins C and A. The seeds are ground into flour resembling that of buckwheat and made into bread.

Bur cucumber

Gourd Family
Sicyos angulatus L.
Cucurbitaceae

Key features: A vine that climbs by tendrils; leaves alternate, large with 3-to-5 angled lobes; flowers bell-shaped, greenish white.

Origin: Native.

Life form: Summer annual vine from a fibrous taproot.

Stems: Light green, ridged, sticky-hairy; to 20 feet long or more.

Leaves: Blades rough-hairy, tips blunt or sharp-pointed, margins slightly toothed, palmately veined; 4 to 8 inches wide.

Flowers: Male and female on separate stalks from leaf axils (male, long-stalked, and female, short-stalked); sepals 5, tips threadlike; petals 5-lobed; anthers clustered together; stigmas 3; July through September.

Fruits: Broad, ovoid, about ½ inch long, covered with barbed prickly bristles; seeds brown with 2 whitish knobs at base, flattened.

Distribution: Moist ground, especially along waterways. Common.

In Kentucky: (IP-mostly), (AP-rare), (ME-rare).

The common name refers to the fruits, which resemble small prickly cucumbers. Some people are sensitive to handling these fruits and can break out in a rash from such contact.

This early summer weed can be aggressive and often covers surrounding vegetation, but dies back in the fall.

Rhombic copperleaf

Spurge Family
Acalypha rhomboidea Raf.
Euphorbiaceae

Key features: Leaves diamond-shaped; leaflike bract at the base of the leaf stalk has 5 to 7 sharp lobes, reminiscent of an out-stretched hand; flowers inconspicuous.

Origin: Native.

Life form: Summer annual herb from a taproot.

Stems: Erect, simple or branched, hairs white, flattened or incurved; 8 to 24 inches.

Leaves: Alternate on stalks more than half as long as the blade; blades ovate to rhombic, veined, margins bluntly toothed, bases tapering, tips pointed; 1 to 2½ inches long.

Flowers: In separate male and female clusters on the same plant; female flowers 1 to 3, hidden in bracts below, sepals 4, minute, green; male above in rounded axillary clusters, sepals 4, yellowish green. July through October.

Fruits: Capsules 3-lobed.

Distribution: Disturbed ground in open woods, fields, roadsides, waterways, cultivated beds, thickets. Common.

In Kentucky: AP, IP, ME.

Similar species: Virginia copperleaf (*Acalypha virginica* L.) is a native summer annual to 2 feet tall. The alternate **leaves are lanceolate**, bluntly toothed to entire and on leaf stalks 1 inch long. Male and female flowers are tiny and develop within the same leaflike bract from the leaf axils. This **bract is lobed with 9 to 15 linear-oblong segments and usually much shorter than subtending leaf stalk.** Cher, Sen, Iroq, Shaw, Chick. Disturbed ground in open woods, fields, thickets, roadsides, waterways, thickets, cultivated beds. Common. July through October. In Kentucky: AP, IP, ME.

The plants of this group are also called waxballs because of the waxy balls of pollen that are on the stamens when the plant is in full bloom. The copperleaf name refers to the purple to copper color of the plant in autumn.

Moonseed

Moonseed Family
Menispermum candense L.
Memispermaceae

Key features: Twining vine without tendrils; leaves shallowly 3-to-7-lobed or angled; leaf stalk attached just above leaf base; fruits poisonous, round, black, in dangling clusters.

Origin: Native.

Life form: Perennial vine, twining.

Stems: Slender, woody, smooth, green; to 12 feet long.

Leaves: Alternate, simple; blades broadly ovate to nearly round, smooth, palmately veined; 3 to 5 inches wide.

Flowers: Produced in loose axillary clusters, male and female flowers on separate plants; sepals 4 to 8; petals 4 to 8, greenish, short; stamens 12 to 24 in male flowers; pistils 2 to 4 in female flowers. June through August.

Fruits: Round, small; seed flattened, moon-shaped and ripening in fall.

Distribution: Moist woods, woodland borders, along waterways, thickets. Common.

In Kentucky: AP, IP, ME.

Most members in this family are found in the tropical rain forests throughout the world. Closely related genera have important economic uses; for example, curare, which is a common name for fish poisons derived from the isoquinoline alkaloids in the fruit, is used clinically today as a muscle relaxant in surgical operations. Such alkaloids are poisonous and fatalities have been reported from persons eating these fruits, which look similar to edible wild grapes.

The common name refers to the seeds that are shaped like a crescent moon.

Crabweed

Mulberry Family
Fatoua villosa (Thunb.) Nakai
Moraceae

Key features: Stems glandular-sticky; leaves alternate, stalked; flowers light green, male and female together in small, dense axillary clusters, usually 1 to 2 per axil.

Origin: eastern Asia.

Life form: Annual herb from a fibrous taproot.

Stems: Erect, light green, hairs granular-sticky, some hooked; to 3 feet tall.

Leaves: Blades ovate, margins bluntly toothed, distinctly veined, upper and lower surface glandular-dotted, bases mostly heart-shaped, sometimes squared, or unequal, tips pointed; to 3 inches long.

Flowers: Male: sepals 4, light green, ovate; petals absent; stamens 4, white, exserted; female: sepals green, 4, lanceolate; petals absent; style reddish purple; all densely glandular-hairy. May through October.

Fruits: Achenes white, oval; seeds tiny.

Distribution: Disturbed ground along roadsides, trails, open woods, woodland edges, cultivated beds. Abundant. First sited in Cherokee in 2006 in Bonnycastle where a few plants were seen along the roadside under a red mulberry tree. Since 2009, it has taken hold in Cherokee and is spreading rapidly into the woodlands; especially rampant in Iroquois.

In Kentucky: IP, ME. Listed as a Moderate Threat by the Kentucky Exotic Pest Plant Council.

This herbaceous plant is an unusual member of the Mulberry Family, which consists mostly of trees—hence the common name mulberry weed.

Crabweed is relatively new to Kentucky. It was first reported in the United States in Louisiana in 1964 by the late Kentucky botanist Dr. John Thieret. This species is troublesome in greenhouses, plant nurseries, and disturbed sites in the eastern and lower midwestern states, where it has spread by seeds transported with horticultural materials. It is now considered a noxious weed in North America.

Helleborine
Orchid Family
Epipactis helleborine (L.) Crantz
Orchidaceae

Key features: Leaves alternate, dull green; flowers pale green to pinkish purple, small, arranged in loose to dense terminal clusters.

Origin: Eurasia, North Africa.

Life form: Perennial herb from rhizomes and fibrous roots.

Stems: Lacking except for the densely short-hairy flowering stalk 1 to 2 feet tall.

Leaves: Three to 10, sessile; blades ovate to lanceolate, strongly veined, margins entire, tips pointed; upper leaves shorter, reduced to narrow, green leaflike bracts; to 3½ inches long.

Flowers: Sepals 3, greenish pink, narrow, midvein green; petals 3, ovate, lateral 2 shorter than sepals, lower petal crimson inside, constricted in middle forming a cup, not spurred; ½ to ¾ inches wide. June through July.

Fruits: Capsule obovoid, 6-nerved, densely to sparsely hairy, about ½ inch long; seeds minute.

Distribution: Dry to moist rocky woods, shady wooded slopes. Rare.

In Kentucky: Collected from 2 counties: Jefferson, Campbell.

This weedy orchid was first discovered in 1986 on Shipping Port Island in the Ohio River near Louisville. Soon after, a "strange flower" was discovered between the marigolds and petunias in a home garden in Campbell County, Kentucky. In 1994, it was found in another garden in Alexandria, Kentucky, where it flowered for three straight summers and then disappeared; in 2000, it reappeared in the same spot.

Helleborine is sporadic in its appearance year to year, and colonies can vary in numbers. The seeds are "dustlike" in size, and one fruit capsule can contain over 1,000 or more. Extremely light, these seeds are easily transported by wind and by water, where they may remain afloat for weeks. It was first discovered in North America in Buffalo, New York, in

1882 and is now found in twenty-eight states and six Canadian provinces. It grows in a variety of habitats, including wooded hillsides, dry rocky slopes, railroad embankments, old graveyards, and gardens. It seems to thrive in sites disturbed by humans.

In June, 2005, six plants were found growing in open rocky woods along Beargrass Creek. Today, the population still exits and the number has increased. Another area with only a few plants was located upstream on a shady wooded slope. These mark the second known location in Jefferson County.

Japanese knotweed
Smartweed Family
Polygonum cuspidatum Siebold & Zucc.
Polygonaceae

Key features: Plant shrub-like with jointed stems; leaves alternate, entire; flowers greenish white, in either erect or drooping clusters.

Origin: Japan, China, Korea, and Asia.

Life form: Perennial herb from long, creeping rhizomes.

Stems: Stout, arched, reddish brown, hollow, smooth; to 6 feet tall.

Leaves: Blades broadly ovate, margins entire, often wavy, heavily veined, bases straight to rounded, tip abruptly pointed; 4 to 6 inches long.

Flowers: In numerous forking clusters in upper leaf axils; plants either all female or all male: male flowers in upright clusters and female drooping; ⅛ inch long. July through October.

Fruits: Achenes triangular, shiny, small, covered with widely winged sepals.

Distribution: Moist open woods, woodland edges, waterways, roadsides. Especially troublesome along the banks of the Ohio River. Being eradicated due to its invasive tendencies.

In Kentucky: AP, IP, (ME-rare). Listed as a Severe Threat by the Kentucky Exotic Pest Council.

This plant, also known as fleeceflower, donkey rhubarb, and Japanese bamboo, was introduced into the United States as an ornamental in the late 1900s from Asia. Because of its "bamboo-like" growth it was used as a landscape screen and for erosion control. Today, it is listed by the World Conservation Union as one of the world's worst invasive species.

The extensive rhizomes are capable of damaging roads, retaining walls, and foundations of buildings. They can also diminish the capacity of channels to carry away floodwaters. It is a major invader of temperate riparian corridors.

Pale dock

Smartweed Family
Rumex altissimus Alph. Wood
Polygonaceae

Key features: Leaves lanceolate, shiny; flowers greenish pink turning reddish brown, in densely packed whorls; fruits heart-shaped.

Origin: Native.

Life form: Perennial herb from a fleshy taproot.

Stems: Erect, often reclining with age, stout, light green with darker lines turning brown, smooth; to 4 feet tall.

Leaves: Alternate, short-stalked above, long-stalked below; blades yellowish green above, midvein prominent, margins entire, not wavy, tips pointed; to 10 inches long; papery sheath at the leaf base and stem turning brown, not fringed at summit.

Flowers: In terminal clusters 4 to 12 inches long; sepals 6, ovate; petals absent; $1/8$ inch wide. May through June.

Fruits: Achenes heart-shaped, 3-angled, membranous, distinctly veined, reddish brown; seed shiny.

Distribution: Moist to wet ground. Locally common at Cherokee; rare elsewhere.

In Kentucky: (IP-Inner Bluegrass), (ME-rare).

Pale dock is a native plant with value to wildlife. It grows in moist ground, where it can form a dense cover providing nesting sites and food for small mammals and songbirds. It can be confused with curly dock and broad-leaf dock, both weedy plants from Europe, but the leaf margins are flat, not wavy, and the upper leaf surface is smooth and shiny. Mature fruits are essential for keying out the *Rumex* species.

Other common names are smooth dock and water dock.

Forest bedstraw

Madder Family
Galium circaezens Michx.
Rubiaceae

Key features: Leaves in whorls of 4, blade broadest near the middle; flowers tiny, greenish white, in 2-forked branches; fruits round, bristly.

Origin: Native.

Life form: Perennial herb from fibrous roots and rhizomes.

Stems: Erect to ascending, mostly simple, very slender, light green, 4-angled; 8 to 18 inches tall.

Leaves: Sessile, blades elliptic to ovate, veins 3, margins smooth, bases tapering, tips with minute abrupt tip; to 1½ inch long.

Flowers: Few, in terminal and upper axillary branches; sepals absent; petals 4; stamens 4, short; styles 2, short; ⅛ inch wide. May through July.

Fruits: Round, 2 parts separating into 1-seeded units; seeds black or brown, covered with hooked white bristles.

Distribution: Moist to dry woods, woodland edges, shady roadsides. Uncommon at Cherokee; common at Iroquois.

In Kentucky: AP, IP, ME.

The fruits of some bedstraws have hooked bristles that cling to the fur of animals and the clothing of humans and are easily transported to new areas by this mode of dispersal.

Common greenbrier
Catbrier Family
Smilax rotundifolia L.
Smilaceae

Key features: A climbing vine with tendrils; stems with numerous green prickles with dark tips; leaves alternate, shiny; flowers greenish yellow, in axillary clusters.

Origin: Native.

Life form: Tough woody vine from long slender underground rhizomes.

Stems: To 20 feet or more, round to 4-angled.

Leaves: Blades ovate to rounded, bright green, margins sometimes rolled inward, bases squared off, rounded, or heart-shaped, tip abruptly pointed; to 5 inches long.

Flowers: Male and female flowers on separate plants in numerous open to dense, rounded clusters; male with 6 tepals and 6 stamens: female with 6 tepals and pistil with 3 thick, spreading stigma lobes. May through June.

Fruits: Berries black or dark blue, with whitish bloom; seeds mostly 2.

Distribution: Dry to moist woods, woodland edges, waterways. Abundant at Iroquois, forming tangled masses in open woods and clearings.

In Kentucky: AP, IP, ME.

Similar species: Bristly greenbrier (*Smilax hispida* Raf.) is a woody climbing vine with green stems **densely covered in slender, black, bristlelike prickles**. The ovate leaves, to 5 inches long, are distinctly veined, with leaf margins entire or finely toothed. Berries are black, bitter, with 1 to 2 seeds. Cher, Sen. Moist woods, especially along Beargrass Creek. Uncommon. May through June. In Kentucky: AP, IP, ME.

Some Native American tribes used the common greenbrier to block the path of pursuers because the sharp prickles easily scratched the skin, causing pain. Often forming impenetrable green masses, the species has other common names such as devil's-clothesline and hell-ropes.

The flowers of bristly greenbrier are said to have an odor similar to a dead rat, and this attracts carrion flies—hence the name carrion-flower.

Clearweed

Nettle Family
Pilea pumila (L.) Gray
Urticaceae

Key features: Stems juicy and nearly translucent; leaves opposite, on slender stalks; flowers tiny, greenish tan, in axillary clusters.

Origin: Native.

Life form: Annual herb from fibrous roots.

Stems: Simple or branched from the base, often reclining; to 2 feet tall.

Leaves: Blades ovate to elliptic, thin, with 3 prominent veins, margins blunt or sharply toothed, bases wedge-shaped to rounded, tips sharp-pointed; 1 to 4 inches long.

Flowers: Clusters of either all male or all female flowers, these mixed together: male with 3 to 4 sepals; stamens 3 to 4; female flowers with 3 unequal sepals. July through September.

Fruits: Achenes ovate, flattened, light brown, mottled, subtended by persistent sepals.

Distribution: Moist to wet woods, shady waterways, roadside ditches. Common to locally abundant especially in moist woods.

In Kentucky: AP, IP.

The Cherokee Indians used this plant as a dermatological aid by rubbing the stem between the toes to stop the itching of "athlete's foot." They also made an infusion that was given to children to reduce excessive hunger.

Traditionally used by the Iroquois Indians, they would squeeze the water from the succulent stem and inhale it to help relieve sinus problems.

The common name refers to the translucent quality of the plant.

Wood-nettle

Nettle Family
Laportea canadensis (L.) Wedd.
Urticaceae

Key features: Plant with stiff, white stinging hairs; leaves alternate, long-stalked; flowers inconspicuous, greenish white, male and female in separate clusters.

Origin: Native.

Life form: Perennial herb with tuberous roots.

Stems: Usually solitary, flexuous, with stinging hairs; to 3½ feet tall.

Leaves: Blades ovate to heart-shaped, veins 3, prominent, margins coarsely toothed, bases rounded to square, tips pointed; 3 to 8 inches long.

Flowers: Male in lower leaf axils and female in loose clusters near top of plant; male flowers greenish, sepals 5, narrow; petals absent; stamen 5; female flowers greenish red, sepals 4, unequal; petals absent; pistil 1. July through September.

Fruits: Flattish, asymmetrical, tiny.

Disribution: Moist to wet woods, waterways. Abundant, especially at Cherokee, where it is the dominant plant in moist shady woods in summer.

In Kentucky: AP, IP.

Wood-nettle was used by the Indians in Quebec to make embroidery thread. It was spun without a spindle and twisted on their knees with the palm of the hand. Their crafts were said to rival any made in France. Baskets, cordage, canvas, and fishing nets were also hand crafted from the rind.

Porcelain-berry

Grape Family

[*Ampelopsis brevipedunculata*
(Maxim.) Trautv.]

Vitaceae

Key features: Woody vine with white
pith; leaves variable from shallowly or
deeply 3-to-5-lobed to heart-shaped;
berries showy, ranging in color from
bright blue, purple, pale lilac, to white.

Origin: Asia.

Life form: Deciduous perennial vine
from a vigorous taproot.

Stems: Bark brown, smooth, not
shredding, dotted; grows to 20 feet or
more in a single growing season.

Leaves: Alternate, margins toothed,
veins below hairy, leaf stalk hairy,
especially at point of attachment to leaf
blade; tendrils coiled, pointed, usually
opposite each leaf; 2 to 4 inches wide.

Flowers: In long-stalked terminal and
axillary clusters, wider than long; sepals
5; petals 5, greenish white, stamens 5,
short. June through August.

Fruits: Berry small, round; seeds 1 to 4,
maturing in September to November.

Distribution: Disturbed ground,
woodland edges, thickets, waterways.
Especially invasive in Cherokee, Seneca,
and Shawnee, where it forms "shrouds"
or tangled masses covering existing
plants. Recently established at Iroquois.

In Kentucky: Collected in 5 counties to
date: Jefferson, Gallatin, Campbell,
Fayette, Madison. Listed as a Severe
Threat by the Kentucky Exotic Pest Plant
Council.

This ornamental vine was introduced
from northeast Asia in 1870s and planted
on estates in the eastern United States.
Since then, it has spread from cultiva-

Invasive plant
Cher, Sen, Iroq, Shaw, Chick

tion, where it can rapidly blanket the surrounding vegetation, especially in open areas of the urban landscape. The plant reproduces both vegetatively from stem and root segments as well as from seeds. Birds and other small mammals are attracted by the colorful, tasty fruits and disperse the seeds in their droppings; waterways also carry the floating seeds downstream to a new location.

The genus name, *Ampelopsis,* is Greek for "vine" and refers to the climbing habit of the plant.

Summer grape

Grape Family
Vitis aestivalis Michx.
Vitaceae

Key features: Vigorous vine with tendrils; stems woody with brown pith; leaves variable from unlobed to shallow or deeply 3-to-5-lobed with reddish cobwebby hairs below.

Origin: Native.

Life form: Woody vine climbing to 20 feet or more.

Stems: Bark on older stems shredding, peeling into long strips; young nonwoody branches are reddish green, smooth or covered with rust-colored, cobwebby hairs.

Leaves: Alternate, on nonwoody branches; blades orbicular to ovate in general outline, distinctly veined, margins coarsely toothed, bases heart-shaped with distinct basal sinus; to 6 inches long.

Flowers: Small, in elongated clusters, opposite a leaf; both perfect and unisexual flowers present; sepals a flat disk; petals 5, greenish yellow, tips fused together, falling early; stamens 5; pistil 1; about $1/8$ inch long. May through July.

Fruits: Berries about $1/4$ inch long, dark bluish purple, round, short-stalked, juicy; seeds 2 to 4, maturing in late summer to fall.

Distribution: Dry to moist open woods, woodland edges, thickets, roadsides, waterways. Common.

In Kentucky: AP, IP, ME.

Also called pigeon grape, this attractive native grape climbs over small shrubs and trees by means of tendrils. It is not used in commercial viticulture, because it does not propagate well through dormant cuttings.

The Cherokee Indians grew a variety of summer grape that was cultivated and used in sacred rituals. As a beverage, it was mixed with pokeberry, sugar, and cornmeal and made into a juice. It is also used to make wild grape jelly.

Vitis is Latin for "grape," and *aestivalis* means "pertaining to summer."

Virginia-creeper

Grape Family
Parthenocissus quinquefolia (L.) Planch.
Vitaceae

Key features: Woody trailing vine; pith white, stems with branched tendrils ending with adhesive tips; leaves alternate, palmately compound.

Origin: Native.

Life form: Vine from a taproot.

Stems: Becoming brown and woody with age, heavily dotted; to 50 feet long or more.

Leaves: Long-stalked, leaflets 5 (3), blades elliptic to ovate, dark green above, paler below, margins toothed mainly beyond middle, tips sharp pointed; to 6 inches long.

Flowers: In terminal and axillary clusters borne under the foliage; bisexual or of single sex only; sepals 5 or absent; petals 5, greenish yellow, curved backward; stamens 5, short, anthers yellow; June through August.

Fruits: Berry bluish black, round, on stalks that turn reddish in the fall; seeds 1 to 4.

Distribution: Moist to dry woods, climbing on trees, over thickets, clearings. Common.

In Kentucky: AP, IP, (ME-rare).

This attractive native vine is valued as a garden ornamental for its vigorous growth, deep red fall color, and blue berries. The berries are eaten by several species of birds, which help spread the plant. White-tailed deer browse on the foliage and stems.

The genus name is from the Greek *parthenos* "virgin" and *kissos* "ivy".

English ivy

Ginseng Family
Hedera helix L.
Araliaceae

Key features: A vine; stems climbing or trailing by sticky aerial and nodal roots; leaves alternate, of 2 types; flowers small, greenish white to yellow.

Origin: Europe, w. Africa, and n. Africa.

Life form: Woody evergreen vine from rhizomes.

Stems: Pale green to reddish, unarmed; climbing or trailing to 50 feet or more.

Leaves: Young plants: blade 3-to-5-lobed, upper surface dark green, shiny, veins white, yellowish green below. Mature plants: blades ovate to rhombic, margins entire, paler green above, bases rounded to squared; 1 to 4 inches wide.

Flowers: Produced in terminal rounded clusters; sepals 5, short; petals 5, fleshy; stamens 5; pistils 5; rarely flowers. September through October.

Fruits: Berry ripening black, ¼ inch wide; seeds 1 to 5, stone-like, poisonous.

Distribution: Disturbed ground in open woods, woodland edges, roadsides, shady turf. The oldest and longest fruiting vines in Cherokee are in areas that border estate houses. Locally invasive.

In Kentucky: (AP-rare), (IP-few counties), (ME-rare). Listed as a Significant Threat by the Kentucky Exotic Pest Plant Council.

Cultivated since ancient times, this ornamental was introduced into North America by the early settlers. Shade tolerant, it is commonly used as a ground cover in urban landscapes and has escaped to nearby woods, where it forms "ivy deserts." It spreads by

Invasive plant
Cher, Sen, Iroq, Shaw, Chick

vegetative means and rarely sets seeds. Plants can remain immature indefinitely, but sunlight triggers more branching with entire leaf forms and occasional flowering. In England, it has been documented that the seeds are spread by house sparrows and black birds. In Kentucky, it has been documented from fifteen counties, but is spreading.

Hedera is from an old generic name meaning "twining."

Wild ginger

Birthwort Family
Asarum canadense L.
Aristolochiaceae

Key features: Plant from thick, creeping rhizomes; leaves opposite, heart-shaped; flowers brownish red, solitary on a short stalk in the fork in the leaves.

Origin: Native.

Life form: Perennial herb.

Stems: Lacking except for the flowering stems.

Leaves: Basal, 2, on stalks branched at ground level rising to 7 inches tall; blades dark silky green, palmate veins distinct, margins smooth, basal sinus rounded, tips rounded to short-pointed; 2½ to 6 inches wide.

Flowers: Small, cup-shaped, close to the ground and often hidden by leaf litter; sepals 3, brownish red, hairy, lobes vary from long-tapering, spreading, or reflexed; petals absent; stamens 12; stigma 6-lobed; ½ inch wide. April though June.

Fruits: Capsules brown; seeds large, ovoid, wrinkled with a fleshy appendage.

Distribution: Moist woods, ravine slopes. Common in Cherokee; rare in Iroquois and Seneca.

In Kentucky: AP, IP, ME.

This mostly tropical and warm-temperate family has many species that are cultivated because of their curious flower shapes and unusual colors.

The thick, creeping rootstock has a strong ginger-like odor and was used by colonists as a substitute for the well-known tropical spice. Some Native Americans roasted the roots and made them into a powder and sprinkled them on clothing like perfume. Others processed the roots with lye water and used the result to season food and take the muddy taste away from fresh fish.

The seeds have fleshy appendages, called elaisomes, that ants are attracted to. They carry the seeds to their underground chambers, where they eat the elaisomes and then discard the seed, aiding in the spread of this species. It is not unusual to find large colonies of this wildflower in woods throughout the eastern United States. By late autumn the leaves die back and disappear.

English plantain
Plantain Family
Plantago lanceolata L.
Plantaginaceae

Key features: Leaves basal, narrow; flowers tiny, golden brown, in short cylindrical clusters at the tips of the stem.

Origin: Eurasia.

Life form: Perennial herb from a fibrous taproot.

Stems: Lacking except for the flowering stalk 8 to 30 inches tall.

Leaves: Stalked, blades narrow, veins 3 to several, prominent, margins entire, tips pointed; 4 to 16 inches long.

Flowers: Sepals 4; petals 4, ovate, translucent, spreading to reflexed; stamens 4, prominent; pistil 1 with long style. April through October.

Fruits: Capsules brown; seeds 2, tiny, dark brown and glossy.

Distribution: Disturbed ground especially in turf, fields, roadsides. Common weed.

In Kentucky: AP, IP, ME.

Similar species: American plantain (*Plantago rugelii* Decne.) is a native perennial herb to 12 inches tall. The **basal leaves** are **broadly ovate to elliptic,** wavy-margined with stalks **reddish purple at base.** The tiny, golden brown flowers are in **dense, elongated terminal clusters.** Cher, Sen, Iroq, Shaw, Chick. Disturbed ground in open woods, turf, fields, roadsides. Common weed. April through October. In Kentucky: AP, IP, ME.

Native Americans called the plantains "the white man's foot" because the plants accompanied the "white man" wherever he traveled in the New World.

The leaves were used by early settlers as a medicine to cure snake and insect bites. The Cherokee Indians made a compound infusion from the leaves that was given to strengthen a child learning to walk.

Crested coral-root
Orchid Family
Hexalectris spicata (Walter) Barnhart
Orchidaceae

Key features: Plant myco-heterotrophic with yellowish brown to purple stems; flowers brownish yellow with bright purplish brown ridges on lower lip, several, in a loose spike.

Origin: Native.

Life form: Perennial herb from a stout, branching, jointed rhizome.

Stems: Erect, smooth; 12 to 26 inches tall.

Leaves: Scalelike, purplish.

Flowers: Outer sepals brownish yellow with brown stripes, recurved; inner lower lip largest, yellow with several purplish brown ridges, recurved, lacking a spur; central column thick, fleshy, cream-colored; 1 inch wide. July through August.

Fruits: Capsule strongly 3-ribbed, hanging.

Distribution: Open disturbed woods over limestone. Rare.

In Kentucky: (AP-rare), IP.

This unique species is totally dependent on fungi attached to the roots of other plants and is often found in oak, pine, or red cedar woods. In Cherokee, it was first spotted in 2008 growing amid invasive ground covers by Chris Bidwell, a noted photographer. Detecting this plant is difficult because it lacks green leaves and the brownish yellow stems and flowers blend in easily with the surrounding vegetation.

The genus name is from the Greek *hex* "six" and *alectryon* "cock," describing the crest of the lip, which alludes to a rooster's crest.

Crane-fly orchid
Orchid Family
Tipularia discolor (Pursh) Nutt.
Orchidaceae

Key features: Plant from a corm; single basal leaf, pleated, light green above with purple spots, dark purple below; flowers many, brownish green, on short stalks.

Origin: Native.

Life form: Perennial herb.

Stems: Lacking except for the flowering stalk to 18 inches tall.

Leaves: Single basal leaf appears in late summer, remains green until late spring, then withers when the flowering stalk appears; blade elliptic to ovate, somewhat heart-shaped, distinctly veined, margins entire, tips pointed; 2 to 4 inches long.

Flowers: Borne in a terminal raceme; outer sepals 3, spreading, elliptic; inner petals 3, with 2 linear side petals, central petal spurred, extending backward; ½ to ¾ inch long. June through August.

Fruits: Capsule ovoid, slightly ridged, less than ½ inch long, hanging parallel to the stem; seeds minute.

Distribution: Moist to dry woods. Common at Iroquois: rare at Cherokee.

In Kentucky: AP, (IP-but rare in Inner Bluegrass), ME.

This unique orchid is often overlooked because the brownish green flowers and stem blend in with the leaf litter. At Iroquois, in some years this orchid is common, making a hike through the rich woods magical.

The Orchid family is very large and distributed throughout the world. Highly sought after for their spectacular flowers, legends surround the early discoveries, importation, and great sums of money obtained at auctions for these prized plants. Although legislation prohibiting collecting, sale, or export of these much sought after rarities is in place, many species are in danger of extinction due to loss of habitat.

Common cat-tail

Cat-tail Family
Typha latifolia L.
Typhaceae

Key features: Plant of wet areas with long, thick rhizomes; leaves long-linear, flat; flowers brown, in densely crowded cylindrical spikes.

Origin: Native.

Life form: Perennial herb.

Stems: Erect, stout, up to 8 feet tall.

Leaves: Erect, light green to bluish green, exceeding the terminal spikes; leaf sheath closely enveloping the lower part of stem.

Flowers: In terminal cylindrical spikes with 2 usually connected portions; male flowers above, stamens 2 to 5, intermixed with white, hairlike bracts, soon falling; female flowers below, green at first, turning brown, long-lasting. June through November.

Fruits: Nutlike, minute, with copious white hairs arising near the base.

Distribution: Moist to wet ground, roadside ditches, pond margins. Uncommon, but can be locally common where found.

In Kentucky: AP, IP, ME.

Cat-tails are a familiar sight in wet areas throughout the state. Its presence is often an indication of disturbance and is among one of the first species to invade areas adversely affected by human activities. Tolerant of a wide range of soil and water conditions, once established this species can spread via the long rhizomes.

The female flowers are persistent, with the fruit forming a dark brown, showy, cigar-like spike up to 7 inches long covered with hairy fruits. These minute, single-seeded fruits are produced in great number, with as many as 70,000 per inflorescence.

This plant has provided many uses to humans throughout the world and has been used in the paper-making industry as well as for mats, chair seats, pillows, and baskets.

Beechdrops

Broomrape Family
Epifagus virginiana (L.) Barton
Orobanchaceae

Key features: A root parasite; stems tan with purplish tan stripes; flowers inconspicuous, cream-colored with purplish brown bands.

Origin: Native.

Life form: Parasitic plant on the roots of beech trees.

Stems: Stiff, single or many, freely branching, ascending, turning dark brown with age; 5 to 18 inches tall.

Leaves: Alternate, small, reduced to scales, triangular-ovate.

Flowers: Alternate, scattered along the stem, of 2 types: lower and middle ones small, bud-like, brown and fertile; upper ones small, sterile, sepals 5, pale with deep purple-tipped teeth; petals tubular, 4-lobed, slightly curved, cream colored with purplish brown bands; about $1/16$ inch long. August through October.

Fruits: Capsule 2-valved at top, rounded, small; seeds many, minute.

Distribution: Beech woods. Common in Cherokee and Iroquois; uncommon elsewhere.

In Kentucky: AP, IP, ME.

This odd plant looks dead even when it is in bloom, as it has no chlorophyll and lacks green coloration. Beechdrops obtains its nutrients from the roots of beech trees, and without them it could not survive. Instead of typical roots, it has a large, swollen organ with clamp-like or sucking structures that penetrate the root of the host plant to obtain nourishment from the parasite.

The genus name is from the Greek *epi* "upon" and *phagos* "beech." Other common names are cancer-root and cancer-drop and have been known in folk medicine as a remedy for dysentery and cancerous ulcers.

Eastern red-cedar

Cypress Family
Juniperus virginiana L.
Cupressaceae

Key features: Evergreen tree; leaves opposite or whorled, needlelike and scalelike; cones berry-like.

Origin: Native.

Life form: Narrow columnar when young to broadly pyramidal at maturity, to 50 feet tall.

Leaves: Of 2 types: young leaves small, narrow, needlelike, spreading; mature leaves scalelike, blue-green, 4-ranked, blunt-tipped, closely appressed, overlapping, needle bases extending downward on stem; to ¼ inch long.

Cones: Scales fleshy, dark blue, covered with a whitish tinge giving them a sky blue color; seeds 1 to 2. "Juniper berries" mature in October through December.

Distribution: Dry upland woods, fields, limestone outcrops, roadsides. Uncommon. A special collection of Eastern red-cedars have been planted in Seneca Park.

In Kentucky: AP, IP, ME.

Economically important, this species is known for its aromatic odor, and the lumber has been used traditionally to make moth-proof cedar chests, closet linings, fence posts, and furniture. It was once the chief wood used to make pencils but has been mostly replaced by a western species named incense-cedar. The oil of cedar, which is distilled from the leaves and wood, is also valuable in the manufacture of polishes, perfumes, and medicines; the cones are used to flavor gin.

Eastern red-cedar is a pioneer species and one of the first trees to invade and become established in disturbed, cleared, or eroded land. During the Dust Bowl drought in the 1930s, farmers throughout the Great Plains were encouraged to plant it in great quantities as shelterbelts or windbreaks because it grew well and could endure the adverse conditions.

White pine

Pine Family
Pinus strobus L.
Pinaceae

Key features: Tall, straight-trunked evergreen tree; needles in bundles of 5; cones cylindrical and dangling.

Origin: Native to eastern North America.

Life form: Evergreen tree to 100 feet with graceful, horizontal, upturned branches spreading to 40 feet wide.

Leaves: Needles in bundles of 5, rarely 3 or 4, bluish green with white stomatic lines on the inner surface, slender, soft and flexible; to 5 inches long.

Cones: Dangling, 4 to 6 inches long, with thick, brown, sticky cone scales; mature in fall of second year.

Distribution: Open areas, dry ridges. Uncommon.

In Kentucky: AP, (IP-disjunct).

One of the most valuable timber trees in the coniferous forests of the northeastern United States, this dominant species was nearly wiped out as a result of ruthless logging, which began in the colonial days in New England and spread westward, hitting the Great Lake states in the 1880s and 1890s. Because it is smooth, strong, and straight-grained, its wood was in demand for building ships, home construction, covered bridges, and railroads.

This species is found in shady ravines with hemlock, especially in Red River Gorge, but scarce elsewhere in the state. However, it is sold in the nursery trade and used in landscape plantings.

Pawpaw

Pawpaw Family
Asimina triloba (L.) Dunal
Annonaceae

Key features: Tropical-looking large shrub or small tree; leaves alternate, light green; flowers pale green, turning brown, and ending in a rich dark wine color; fruits "banana-like."

Origin: Native.

Life form: Small single- to multi-stemmed, coarsely branched, to 25 feet, often forming colonies.

Leaves: Blades obovate-oblong, lower midvein raised with scattered golden hairs, margins entire, narrowed at base to a short stalk, tips often abruptly pointed; up to 14 inches long; scent of green pepper when crushed.

Flowers: Solitary, nodding; sepals 3, broadly ovate, hairy; petals 6 with 3 small inner ones and 3 larger outer ones, veins distinct; stamens yellow, numerous, surrounding several green styles; 1 to 2 inches long. March through May.

Fruits: Oblong to rounded, fleshy, green, turning yellowish green to black and wrinkled when ripe, 3 to 6 inches long; seeds several large, dark, flattened.

Distribution: Moist wooded slopes, along waterways. Common at Cherokee and Iroquois; uncommon elsewhere.

In Kentucky: AP, IP, ME.

This beautiful shrub or understory tree often grows in association with maple-basswood and maple-beech woods. It is the only representative in Kentucky of a large tropical family that is valued for its fruits, such as cherimoya, sweet sop, sour sop, and custard apple. The overall appearance has a "tropical" look with the large green leaves and unique yellow-orange, banana-shaped fruits that are soft, custard-like, and tasty and are sought after by small mammals. It is sometimes locally called Kentucky banana.

The first reference of this tree occurs in the chronicles of DeSoto's expedition exploring the Mississippi Valley in 1541. It is said that the edible fruits, which are rich in vitamins A and C, helped keep the conquistadores from starvation.

Another common name is fetidshrub, describing the foul odor of the maroon flowers, which resemble rotting meat.

Ironwood

Birch Family
Carpinus caroliniana Walter
Betulaceae

Key features: Tree with smooth, "muscular" blue-gray bark; leaves alternate, margins sharply doubly toothed; fruits with a 3-lobed leaflike bract.

Origin: Native.

Life form: Small multi-stemmed shrub or single-stemmed tree to 30 feet tall.

Leaves: Blades ovate to elliptical, dark green, glossy, smooth above, hairs on veins below, lateral veins not forked, bases rounded, tips pointed; wrinkled when they first appear; 1 to 4 inches long.

Flowers: Male and female on the same plant: male at the ends of short lateral branches and female in terminal catkins. April through May.

Fruits: In loose, drooping clusters; nutlets small, ovoid, ribbed, attached to a light green, distinctly veined bract.

Distribution: Moist woods, woodland edges, waterways. Uncommon.

In Kentucky: AP, IP, ME.

This unique native tree is also called blue beech and water beech because of the strong wood. In the past, charcoal made from this species was used in the manufacturing of gunpowder. The hanging fruit clusters with spreading leafy bracts are distinctive.

Hop-hornbeam

Birch Family
Ostrya virginiana (Mill) K.Koch
Betulaceae

Key features: Tree with grayish brown bark broken into narrow, loose strips; leaves alternate, doubly toothed; fruits with inflated "hop-like" papery bract.

Origin: Native.

Life form: Small to medium understory tree to 30 feet tall, occasionally taller with many horizontal or drooping branches. Often produces understory seedlings and saplings.

Leaves: Blades oblong to ovate, dark green above, paler below, soft-hairy, lateral veins forked, bases rounded to slightly heart-shaped, tips pointed; 2 to 5 inches long.

Flowers: Male and female flowers on same plant: male in slender hanging catkins, appearing with leaves; female catkins shorter, upright. April through May.

Fruits: In drooping clusters 1½ to 2 inches long; nutlets small, ovoid, flattened, enclosed in an inflated papery bract.

Distribution: Dry to moist woods, especially in uplands. Uncommon.

In Kentucky: AP, IP, ME.

This graceful native tree, known for its very strong, hard wood used for making tool handles and fence posts, is also called ironwood, leverwood, and deerwood.

The generic name is derived from the ancient Greek name of a tree with hard wood.

The Cherokee Indians made an infusion from the bark that was taken to build up the blood, relieve toothaches and sore muscles, and as a treatment for rectal cancer.

The common name hop-hornbeam refers to the resemblance of the overlapping fruits to that of hops.

Persimmon

Ebony Family
Diospyros virginiana L.
Ebenaceae

Key features: Bark almost black, deeply ridged, and blocky; leaves alternate, dark green above, smooth; fruits round, yellow, ripening to bright orange when mature.

Origin: Native.

Life form: Medium-size to large-size tree to 80 feet tall.

Leaves: Short-stalked; blades oblong-ovate, smooth, midvein above yellow, margins entire, bases rounded, some unequal, tips pointed; to 6 inches long, but variable in size on the same twig.

Flowers: Male and female on separate trees or sometimes mixed: male in branched clusters and female solitary in leaf axils. May through June.

Fruits: Berry 1 to 2 inches wide, juicy, tipped with a persistent style, sepals green, 4-lobed, persistent at base; seeds up to 8, flat.

Distribution: Dry to moist woods, woodland edges, fields, roadsides. Common at Iroquois; uncommon to rare elsewhere.

In Kentucky: AP, IP, ME.

The genus name, *Diospyros* "fruit of the gods," aptly describes the sweet, juicy, edible fruits that ripen in late fall. They can remain on the tree for a long time, but once picked, they spoil quickly. An old saying goes that if you shake a tree and the ripe fruits fall easily to the ground, then they are perfect for eating, and if not, they are sour and will make your mouth pucker. The fruit is used in making puddings, breads, pies, wines, and preserves.

Holes drilled through the seeds were used as buttons during the Civil War during times of shortages. The wood, known as the "ebony of America," is heavy, strong, and dark.

Redbud

Legume Family
Cercis canadensis L.
Fabaceae

Key features: Leaves alternate, heart-shaped; flowers pealike, pinkish purple, in clusters borne on the branches and on the trunk; fruit pods flattened.

Origin: Native.

Life form: Small understory tree to 30 feet tall with slender, zigzagged stems; trunk often splits near base, becoming multi-stemmed.

Leaves: Blade dark green above, paler below, veins distinct, margins entire, tip abruptly pointed; 3 to 6 inches long.

Flowers: Small, in clusters of 4 to 8 on threadlike stalks; opening before the leaves. March through May.

Fruits: Pods slender, turning brown at maturity, 2 to 4 inches long; seeds flat, light brown.

Distribution: Moist to rocky woods, woodland edges, roadsides, especially in limestone. Common.

In Kentucky: AP, IP, ME.

Redbud is often called Judas tree, but this name refers to the species, *Cercis siliquastrum,* which is native to southern Europe and western Asia. According to legend, Judas hung himself on this tree after betraying Christ—hence the rosy-colored flowers.

This tree is a welcome sign of spring. Placing the flowering branches inside the home was said to "drive winter out."

American beech
Beech Family
Fagus grandifolia Ehrh.
Fagaceae

Key features: Large tree with smooth, gray bark; leaves alternate, ovate, toothed; fruits a 4-valved bur with recurved prickles.

Origin: Native.

Life form: Large tree to 75 (100) feet with a shallow root system.

Leaves: Leaf stalks silky-hairy, blades dark green above, pale greenish yellow below, margins toothed, veins straight running to each tooth, bases wedge-shaped to rounded, tips tapering; 2 to 4 inches long.

Flowers: Male and female flowers on same tree: male dangling on slender stalks, heads round with 8 to 12 stamens; female in pairs at the ends of the new shoots. March through May.

Fruits: Nuts small, rusty brown; 3/4 to 1 inch long.

Distribution: Moist shady woods, ravine slopes, open turf. Common at Iroquois; uncommon elsewhere.

In Kentucky: AP, IP, ME.

The fruit of this majestic tree was a nutritious food source used by the Native Americans and early settlers. The Iroquois Indians crushed the nuts and mixed them with cornmeal, beans, or berries and made bread. Nutmeat oil was used as a hair conditioner, either alone or with bear grease. Other tribes relied on beech nuts for food and as a remedy to expel worms. The nuts were also used for buttons.

American beech is an important component of mesophytic forests of the eastern United States, where they occur in moist, well-drained soils in the beech-maple forests and the northern hardwood forests of the Great Lakes, the northeast, and in higher elevations in the Appalachian Mountains.

Today, the deadly Beech bark disease is threatening the health of these beautiful trees, but it has not been reported in Kentucky.

White oak
Beech Family
Quercus alba L.
Fagaceae

Key features: Tree large, with light gray, scaly bark; leaves alternate, lobes 5 to 10, fingerlike; acorn with cap covering less than ¹/₂ of the nut.

Origin: Native.

Life form: Tree up to 100 feet tall or more, the crown wide-spreading in the open, more upright and narrow in woods.

Leaves: Short-stalked, blade obovate in general outline, pinnately lobed, rounded, ascending, sinus depth varies from shallow to deeply cut, upper surface bright green, not glossy, lower pale, bases narrowing to wedge-shaped base; 3¹/₂ inches to 8 inches long.

Flowers: Male and female on same tree but separate: male flowers tiny, yellowish green in slender, drooping catkins, 2 to 4 inches long; female flowers tiny, inconspicuous, reddish brown, solitary or in small clusters. April through May.

Fruits: Acorns sessile to short-stalked, 1 inch long, light brown to tan, shiny, cap blunt and knobby.

Distribution: Moist to dry upland woods and ridges, especially common at Iroquois, where it is associated with chestnut oak.

In Kentucky: AP, IP, ME.

Similar species: **Post oak** (*Quercus stellata* Wangenh.) is a native oak to 60 feet tall. The alternate leaves are dark green above and paler below with star-shaped hairs. They are **usually 5-lobed, with the opposite middle, squared ones largest and forming a "cross."** The short-stalked acorns are solitary or in pairs with bowl-shaped cups with downy scales **covering ¹/₂ of nut.** In dry, upland woods, and ridge tops in poorer soil. It is associated with black jack oak, especially on the northern ridges at Iroquois. Uncommon. In Kentucky: AP, IP, ME.

White oak is a magnificent tree that is one of the dominant species in mixed oak forests and reported to have a life span of 500 to 800 years. The strong, heavy wood is close-grained and considered to be one of the most valuable of lumber trees. It is used for making whiskey barrels, furniture, lumber, and railroad ties. The edible acorns are rich in fat and were made into flour, while the leaves were used to wrap dough for bread making.

Post oak is also known as box white oak and iron oak, which describes the strong, heavy, hard wood.

Bur oak

Beech Family

Quercus macrocarpa Michx.
Fagaceae

Key features: Large, majestic tree with stout, often corky branches; leaves 5-to-9-lobed, broadest above central deep sinus; acorn caps distinctly fringed.

Origin: Native.

Life form: Tree to 150 feet with broad-spreading canopy and with deeply furrowed bark.

Leaves: Alternate, obovate in general outline, blades firm, lobes rounded, upper surface shiny dark green, lower pale green to silvery, hairy, margins entire; 6 to 10 inches long.

Flowers: Male and female on the same tree: male in slender catkins and female inconspicuous, singly or in clusters at the ends of branches. April through May.

Fruits: Nut light brown, elongated, 2 inches long, cap covering most of nut.

Distribution: Open fields; rare at Iroquois and Shawnee; uncommon at Cherokee.

In Kentucky: (IP-mostly Bluegrass Region), (ME-rare).

Living 200 to 300 years or more, this majestic, spectacular oak is a common tree growing in the ancient woodland-pastures of the central Bluegrass. Standing alone in fields or bottomlands, it towered over the understory of native shrubs and grasses. Easily recognized by the massive trunk and rugged branches, the acorns are the largest of any North American oak and produce a good seed crop every two to three years; these are an important food source for wildlife. It is often used in urban landscapes because it is drought tolerant and sturdy.

Blackjack oak

Beech Family
Quercus marilandica Munchh.
Fagaceae

Key features: Tree with gnarly, twisted branches forming an irregular crown; leaves alternate, 3-lobed, prickly; acorns with loose, hairy scales on cap.

Origin: Native.

Life form: Small to medium-size deciduous tree to 30 feet, usually with a short trunk.

Leaves: Blade base rounded, narrow, flaring upward into a broad apex, lobes 3, shallow, glossy dark green above, pale with reddish brown hairs below (these easily rub off); 5 to 7 inches long.

Flowers: Male and female on the same tree: male in hanging catkins, 2 to 4 inches long; female small, single or paired. April through May.

Fruits: Acorns, 1 to 2 on short stalk; cap covers about ½ of nut.

Distribution: Uplands in poor, dry soil, especially on northwest ridges and slopes. Uncommon.

In Kentucky: AP, (IP-absent from Inner Bluegrass), (ME-rare).

This distinct, gnarly tree with almost black, deeply furrowed bark with orange fissures is found growing on upland ridges at Iroquois in poor, dry soil. The presence of this species is a good indicator of poor soil throughout its range from New York, west to Nebraska, south to Texas, and Florida.

The acorns are valuable food for wild turkeys and white-tailed deer.

Chestnut oak

Beech Family
Quercus montana Willd.
Fagaceae
(*syn=Quercus prinus* L.)

Key features: Tree with dark, deep V-shaped furrows in mature bark; leaves alternate, margins with rounded teeth, usually lacking glands; acorns large, nut dark brown.

Origin: Native.

Habit: Medium-size tree; old trees get very wide at girth, but are not much taller than 70 feet.

Leaves: Blades obovate to elliptical, upper leaf surface dark yellowish green, smooth, thick, paler green below, hairs asymmetrical with tufts of longer hairs on midvein, bases wedge-shaped, tips bluntly pointed; 4 to 8 inches long.

Flowers: Male and female on same tree: male in slender drooping catkins and female solitary or in small spikes. April through May.

Fruits: Acorns up to 1¾ inches long, short-stalked, cap bowl-shaped, sharp-rimmed, warty, covering ⅓ of the deep chestnut brown, oblong nut.

Distribution: Dry rocky uplands and ridges on shallow acidic soils. Common in Iroquois; rare in Seneca.

In Kentucky: (AP-common), (IP-scattered).

Similar species: Chinquapin oak (*Quercus muhlenbergii* Engelm.) is a native oak growing to 80 feet tall. It has **gray to yellowish, thin, papery bark** and alternate leaves that are obovate, **shallowly lobed or toothed, the incurved teeth ending in a gland or callus at the point.** The 1 to 2 acorns are rounded, **less than ¾ inch long, light brown** with a bowl-like cap that covers about ¼ to ½ of the nut. Cher, Sen, Iroq. Moist to dry upland woods, on ledges in limestone soils along riparian corridors. Uncommon. April through May. In Kentucky: AP, IP, ME.

With its deeply V-shaped furrowed bark when mature, Chestnut oak is perhaps one of the most spectacular oaks at Iroquois. Growing in shallow soils on dry, rocky south- and west-facing slopes, this species is also called rock oak and rock chestnut oak. The leaves decompose slowly, and leaf litter on the ground around stands of this species is respon-sible for the sparse herbaceous vegetation. Low-bush blueberry is found in association with this oak, but very little else.

Northern red oak

Beech Family
Quercus rubra L.
Fagaceae

Key features: Bark with broad silvery strips; leaves alternate, lobes pinnate, forward-pointing, bristle-tipped, shallow sinuses cut less than ¹/₂ distance to midrib; buds and twigs red-brown, rounded; cap scales acute.

Origin: Native.

Life form: A large deciduous tree to 100 feet tall with a rounded, symmetrical crown and straight trunk.

Leaves: Blades elliptic to obovate, lobes 7 to 11, margins irregularly toothed, upper surface dull green, smooth, often reddish-tinged, lower with axillary tufts of hairs in veins; 5 to 8 inches long.

Flowers: Male and female on same tree: male in loose, drooping catkins; female solitary, sessile or short-stalked. April through May.

Fruits: Acorns solitary or paired; nut oblong to broadly ovoid, chestnut-brown with grayish streaks, saucerlike cap enclosing less than ¹/₂ of nut.

Distribution: Moist woods. Common at Iroquois and Cherokee; uncommon elsewhere.

In Kentucky: AP, IP, ME.

Similar species: Black oak (*Quercus velutina* Lam.) is a stately native, deciduous tree with **brownish black bark** that is **deeply furrowed and blocky with a yellowish orange inner bark.** The leaves, 5 to 10 inches long, are ovate to obovate in general outline with 5 to 9 lobes **separated by deep sinuses**, lustrous shiny green above and with hairs that rub off below. Acorn cap covers more than ¹/₂ of the ovoid nut. Note the **buds are squarish, gray, and hairy.** Cher, Sen, Iroq, Shaw. Moist uplands. Common in Cherokee and Iroquois; uncommon elsewhere. In Kentucky: AP, IP, ME.

Similar species: Shumard oak (*Quercus shumardii* Buckley) is a native, deciduous tree especially common in the Bluegrass Region. The alternate, simple leaves are elliptic to obovate with 5 to 9 **lobes cut more than half way to the midrib and closing toward the**

apex. The upper surface is shiny, dark green, and the lower with large brown tufts in the vein axils. Acorn cap scales with pale margins and tips covering ¼ or less of nut. Note the **buds are 5-angled and both the buds and twigs are grayish green.** Cher, Iroq, Shaw. Moist woods. Uncommon. In Kentucky: (AP-rare), (IP-mostly in Bluegrass Region), ME.

These species make up an important component of the eastern deciduous forests. All three are known to hybridize, sometimes making identification difficult.

Sweetgum

Witchhazel Family
Liquidambar styraciflua L.
Hamamelidaceae

Key features: Twigs corky ridged; leaves star-shaped with 5 to 7 cut lobes; fruit in dangling, rounded heads with 2-beaked projections sticking out in every direction.

Origin: Native.

Life form: Large tree to 80 feet with gray bark that has narrow, flaky ridges.

Leaves: Blades palmately lobed, shiny dark green above, paler below, with axillary tufts of rusty-colored hairs in veins, margins finely toothed, tips pointed; 4 to 7 inches long.

Flowers: Male and female flowers on the same plant: male flowers in upright clusters 3 to 4 inches long and female flowers in heads on long stalks. May through June.

Fruits: Capsules round, prickly, 1 to 2 inches wide, glossy green at first, turning brown; seeds winged.

Distribution: Floodplains, waterways, moist woods. Common at Iroquois in low moist woods; uncommon elsewhere.

In Kentucky: AP, IP, ME.

Sweetgum was the first plant in the Witchhazel Family to be introduced into England from the New World. The genus only contains one species in North America. The cut wood exudes a fragrant liquid or resin from the cracks of the bark—hence the genus name, *Liquidambar,* which is derived from the Latin *liquidus* "fluid" and *amber,* Arabic for the fragrant resin.

In Appalachia, the twigs were soaked in water or brandy, chewed, and used to clean teeth.

Sassafras

Laurel Family
Sassafras albidum (Nutt.) Nees.
Lauraceae

Key features: Bark reddish brown, furrowed; leaves alternate, entire or with 2 or 3 lobes, aromatic when rubbed or crushed; fruits blue on dark red stalks.

Origin: Native.

Life form: Small to medium-size tree with upturned pale green to olive twigs.

Leaves: Simple, short-stalked; blades bright green above, paler below, margins entire, bases variable from sharply tapered, rounded, or wedge-shaped, tips blunt; 3 to 6 inches long.

Flowers: Male and female on different plants; inconspicuous, yellow, clustered on drooping stalks 2 inches long, appearing before or with young leaves. April through May.

Fruits: Fleshy, ovoid, to ½ inch long on dark red stalks with cuplike tips.

Distribution: Woodland edges, road-sides, fields. Uncommon.

In Kentucky: AP, IP, ME.

This tree is beautiful all year round; the small clusters of yellow flowers appear in spring before the dark green leaves, which turn various shades of yellow, orange, or reddish purple in fall. Mature trees, such as those growing along the road at Summit Field in Iroquois, are magnificent, with distinct reddish brown, furrowed bark. Birds readily consume the oily fruits.

The early settlers boiled the roots with molasses and made an extract that was used in fermenting flavorful beer and in making root beer.

Tuliptree

Magnolia Family
Liriodendron tulipifera L.
Magnoliaceae

Key features: Tree to 100 feet tall or more; leaves alternate, square or tulip-like in general outline; flowers showy, greenish yellow with orange spots at base; fruits upright, cone-like.

Origin: Native.

Life form: Grows with a straight trunk devoid of lower branches in the woods and a narrow canopy.

Leaves: Blades 4-lobed with a wide notch at summit, upper surface dark green, shiny, lower pale, central vein ends at notch, bases rounded to squared, tips short-pointed; 3 to 7 inches wide.

Flowers: Sepals 3, green, reflexed; petals 6, in 2 rows; anthers numerous, spreading outward; pistils flat, narrow, overlapping, forming a "cone." May through June.

Fruits: Dry, cone-like structure 2 to 3 inches long that breaks into samaras with 1 to 2 seeds; distinct brown upward tapered scales are left in winter.

Distribution: Rich moist low woods, slopes. Common.

In Kentucky: AP, IP, ME.

Some of the most magnificent tuliptrees are found growing in the low moist woods along with American beech in the southern and southwestern portions of Iroquois.

The genus name is from the Greek *lirion,* "lily" or "tulip," and *dendron* "tree." This beautiful native tree is also called tulip magnolia, yellow poplar, white poplar, and tulip poplar.

In 1956, the Kentucky General Assembly first ruled on the issue of having an official state tree and tuliptree won over the other two candidates, catalpa and sycamore. However, it was not properly documented and lost its state title to the Kentucky coffee tree. After years of heated arguments on this subject, tuliptree was again reinstated and officially named the state tree of Kentucky on March 9, 1994, and the Kentucky coffee tree was designated as the State Heritage tree.

The straight trunks were made into dugout canoes by Native Americans. Today, the wood is a valuable timber and used in manufacturing many wooden products.

Osage-orange
Mulberry Family
Maclura pomifera (Raf.) Schneid.
Moraceae

Key features: Bark orange-brown, furrowed; leaves alternate, long-stalked, usually with stout axillary spines; fruits a large greenish yellow bumpy-surfaced ball, grapefruit-size.

Origin: Native in north Texas, southeast Oklahoma, and Arkansas, spread elsewhere from extensive hedge plantings.

Life form: Large shrub or small round-topped tree with twisted, interlacing branches; often spreading by root-shoots; roots bright orange.

Leaves: Blades ovate to oblong-lanceolate, glossy green above, margins smooth, bases rounded, tips tapered; 3 to 8 inches long.

Flowers: Male and female flowers inconspicuous on separate trees: male in loose short racemes and female in dense rounded heads. April through May.

Fruits: A large ball, to 6 inches wide; seeds small, up to 200, buried in fleshy white latex-filled pulp.

Distribution: Edge of woods, thickets, fields, roadsides. Common.

In Kentucky: IP, ME.

Osage-orange, also known as hedge-apple, was at one time limited to its native range along the Red River Valley. Before the invention of barbed-wire fences in the 1880s, thousands of miles of osage-orange hedges were planted along farm boundaries. Early settlers pruned back the plant to promote bushy growth. It was also used for bows, a common name being bois d'arc in French and bowwood in English. The wood is very strong and resists decay, so it was used for fence posts. It is also one of the hottest-burning fuel woods. Today, it is a common site on the Great Plains and beyond, where it has escaped and naturalized.

The common name comes from the Osage Indian tribe, who claimed the aroma of the late-summer ripe fruit was similar to that of an orange peel. The genus is named in honor of the early American geologist William Maclure (1763–1840).

White mulberry

Mulberry Family
Morus alba L.
Moraceae

Key features: A medium-size tree with a bushy habit of growth; leaves alternate variable, glossy dark green above, few hairs on veins below; fruits are blackberry-like, somewhat bland.

Origin: China.

Life form: Small to medium-size tree to 30 feet tall; with milky sap.

Leaves: Stalked, blades ovate, entire or irregularly 2-to-3-lobed, underside with sparse, appressed hairs mostly along major veins, bases often unequal, tips rounded or pointed; 2 to 4 inches long.

Flowers: Male and female on different plants: male in hanging catkins and female tiny, in compact clusters. April through May.

Fruits: Berries red, turning black, purple, or whitish when mature, about 1 inch long.

Distribution: Disturbed open woods, woodland edges, roadsides, thickets, waterways. Common. Trees are being cut down due to invasive tendencies.

In Kentucky: AP, IP, ME. Listed as a Significant Threat by the Kentucky Exotic Pest Plant Council.

Similar species: Red mulberry (*Morus rubra* L.) is a **native** understory tree to 40 feet tall with a **wide, rounded crown** and twigs that exude a milky sap when cut. The 3-to-8-inch leaf is **ovate to irregularly narrow-lobed, rough above** and with **hairs dense and felty below.** The margins are toothed and the leaf tip is abruptly pointed. The **edible, sweet, juicy berries** are cylindrical, blackberry-like, and about 1 inch long. Cher, Sen, Iroq, Chick. Moist woods, ravine slopes, woodland edges, waterways. Uncommon. April through May. In Kentucky: AP, IP, ME.

White mulberry is an ancient crop plant used as food for the silkworm industry in China. Imported into other parts of the world for developing the silkworm industry, the trees escaped and are now naturalized in Asia, Europe, and America. The tasteless fruits are bland but have high sugar content. Once dried, they were ground into flour and added to dough, becoming a valuable food source and sweetening agent in the diet of the ancient peoples of the mountains of central Asia.

The ripe fruits of the red mulberry are sweet and tasty and are devoured by birds and wildlife. This native understory tree is in decline in our region, possibly due to a bacterial disease or a fungal canker. The fruits are host to mulberry popcorn disease.

Both species are highly variable and it has been documented that both red and white mulberry hybridize making identification difficult.

Sycamore
Sycamore Family
Platanus occidentalis L.
Platanaceae

Key features: Tree with white, peeling, mottled bark; leaves alternate, broadly ovate obscurely 3- to-7-lobed, toothed; fruits ball-shaped that dangle from long stalks.

Origin: Native.

Life form: A large, stately tree to 130 feet tall.

Leaves: Leaf stalk swollen and hollow at base covering a bud; blades palmately lobed, bright green above, smooth, veins white-woolly below, sinuses wide, shallow, bases rounded or squared; 4 to 8 inches wide.

Flowers: Male and female flowers on same tree but in separate clusters: male in leaf axils and female on long terminal stalks. April through May.

Fruits: Heads rounded, about 1 inch wide; made up of numerous 1-seeded achenes with woolly tufts of hairs; fruit persists through winter and breaks up in spring when the downy seeds can be seen floating in the wind.

Distribution: Waterways, low wet to moist woods, disturbed ground. Common. Especially large trees are found growing along the Ohio River at Shawnee. Young saplings have been planted along Beargrass Creek to help stabilize creek bank.

In Kentucky: AP, IP, ME.

Also known as buttonball-tree, buttonwood, plane tree, and white wood, this beautiful tree with mottled gray, tan, and white bark dates back to the Cretaceous time period, where fossil remains have shown that because the climate was much warmer, it once grew in the present-day arctic regions.

Native Americans created an infusion made from mixing three chips taken from the east side of the trunk on this species and the native honeylocust into a drink that was used to cure colds and as a gargle for hoarseness and sore throat.

Unfortunately, a fungal disease—sycamore anthracnose—has affected many trees in the central part of the state. The trees decline and die before reaching maturity.

Downy serviceberry

Rose Family
Amelanchier arborea (F. Michx.) Fernald
Rosaceae

Key features: Shrub or understory tree; leaves alternate, margins sharply toothed, callus-tipped; flowers white, opening before or with young leaves.

Origin: Native.

Life form: Small, airy, 15 to 25 feet tall with warty twigs.

Leaves: Stalked, blades oval to oblong, upper surface bright green, paler below, slightly hairy, bases rounded to heart-shaped, tips pointed; 1½ to 3 inches long.

Flowers: In loose terminal upright and dangling clusters; sepals 5, broadly triangular, soon becoming reflexed; petals 5, narrow; stamens many; styles 5, united below; about 1 inch long. March through April.

Fruits: Plump, berry-like, green at first with whitish cast, turning red to purplish red when ripe in June; borne in small, drooping clusters.

Distribution: Dry to moist upland woods, ravine slopes, waterways. Especially common at Iroquois, where it is an understory tree in upland chestnut-oak woods. Rare elsewhere.

In Kentucky: AP, (IP-but rare in extreme north-central counties), ME.

This attractive native tree is a popular garden plant in the United States and was introduced into England in 1764. The delicate white flowers are a welcome sight in early spring and come into bloom just before the leaves appear in

March or April. It is also known as Juneberry, serviceberry, shadblow, and shadbush. According to tradition, early settlers in New England noted that its flowering coincided with the shad run in the creeks and rivers.

The tasteless fruits are an important food source for birds and mammals.

Wild black cherry

Rose Family
Prunus serotina Ehrh.
Rosaceae

Key features: Tree with reddish brown, glossy bark and branches with many lenticels; leaves alternate, narrow, shiny; flowers white, in dangling cylindrical clusters.

Origin: Native.

Life form: Medium to large tree to 100 feet tall.

Leaves: Short-stalked with pair of reddish glands at summit of leaf stalk; blades lanceolate to oblong, upper surface dark green, lower pale green, margins minutely blunt-toothed with callus tip, bases rounded to wedge-shaped, tips long- or short-pointed; 2 to 5 inches long.

Flowers: Many, small, in clusters; sepals 5; petals 5; stamens numerous, yellow; pistil 1. April through May.

Fruits: Round, dark red, becoming purplish black, juicy, to ½ inch wide; seed 1, ripening in August or September.

Distribution: Moist woods, woodland edges, fields, roadsides. Common.

In Kentucky: AP, IP, ME.

There are two other common names for this tree that are very descriptive: wild black rum and cabinet cherry. The fruits have a bittersweet, wine-like flavor and were formerly used to flavor rum and brandy. The hard, close-grained wood is still used for interior finish, veneer, and to make furniture.

Somewhat weedy, this tree is found growing along the edge of pastures and is poisonous to livestock. It contains prussic acid, which can be fatally toxic to cattle eating the young leaves and shoots and the wilted leaves from mature branches. Symptoms such as convulsions, staggers, bloating, and difficulty breathing are observed before death.

Eastern cottonwood

Willow Family
Populus deltoides W. Bartram ex Marshall
Salicaceae

Key features: Tree with gray, deeply furrowed bark; leaves alternate, leaf stalks flattened; seeds with white "cottony" tufts.

Origin: Native.

Life form: Large tree to 100 feet with open, spreading crown, often with massive spreading branches.

Leaves: Blades broadly triangular-ovate to heart-shaped, dark green above, paler below, margins short-hairy, coarsely toothed below with hard incurved tips, becoming smaller towards tip, bases slightly heart-shaped to squared with 2 or 3 basal glands; 3 to 6 inches long.

Flowers: Male and female on separate trees in dangling catkins appearing before the leaves. March through May.

Fruits: Capsule 3- or 4-valved, splitting open; seeds "cottony."

Distribution: Along waterways, low moist woods. Common, especially along the Ohio River, where many spectacular trees grow.

In Kentucky: (AP-rare), IP, ME.

Also known as poplar, whitewood, and Carolina poplar, this massive tree has light, weak wood, and its limbs tend to break in storms. However, west of the Missouri River, early pioneers found that this species withstood winter blizzards and summer droughts. They made fences, corncribs, stables, and their homes from the wood. Today, this species is grown in tree plantations in major river bottoms as a short-rotation crop. It is the fastest-growing North American tree, with reports of it reaching 100 feet in nine years on rich soils in the South. It is an important source of pulpwood.

The age of the eastern cottonwood is short, usually to 75 years old, and often the heartwood decays and becomes hollow, making for the perfect home for woodpeckers, sapsuckers, owls, and other birds.

Eastern cottonwood sheds copious amounts of pollen and can cause hay fever in sensitive persons. In mid-spring, large quantities of the white fluffy seeds can be seen floating in the wind.

Sandbar willow
Willow Family
Salix exigua Nutt.
Salicaceae
(*syn=Salix interior* Rowlee)

Key features: Multi-stemmed woody shrub; blades narrow, tapering at both ends, hairy below; flowers in catkins, male and female in separate plants.

Origin: Native.

Life form: Densely clumped deciduous shrub to 25 feet tall; often forming thickets from rhizomes and basal shoots.

Leaves: Alternate, simple, sessile or very short-stalked; blades linear-lanceolate, shiny green above, dull green below, midvein distinct, margins entire or with tiny teeth; 2 to 4 inches long.

Flowers: Each flower subtended by a cupped, light green ovate leaflike bract with long silky hairs. March through August.

Fruits: Capsules in clusters; seeds dark brown, tiny, numerous, enclosed in dense tufts of silky white hairs (light brown walls of seed capsules curl back after seeds dispersed).

Distribution: Wet to moist ground, along pond margins, waterways, swamps. Uncommon.

In Kentucky: (AP-rare), IP, ME.

This graceful, leafy-branched willow is also called narrowleaf willow and coyote willow. Intolerant of shade, it is a pioneer species that thrives on sites that are regularly flooded. The seed coats are thin and can germinate in twenty-four hours after being dispersed. Seeds older than one week rarely germinate, so it is important to land in flooded areas. Once established, sandbar willow is an important food source and shelter for wildlife. The rhizomes and basal shoots help stabilize bands from erosion.

Like many willows, this species contains *salicin,* which is closely related chemically to *acetylsalicylic acid,* or aspirin. Native Americans used the many species to treat a variety of ailments, including toothaches, dysentery, stomach aches, and fevers.

Basswood

Basswood Family
Tilia americana L. var. *americana*
Tiliaceae

Key features: Tree with downward spreading branches; leaves alternate, heart-shaped to broadly ovate; floral bract light green, narrow.

Origin: Native.

Life form: A large tree from 60 to 100 feet tall often with several limb sprouts from the base of trunk.

Leaves: Blades with upper surface dark green, smooth, lower paler green, hairs variable but mostly with rusty brown tufts of hairs in axils of veins, margins sharply toothed, bases uneven, heart-shaped to squared, tips abruptly pointed; 4 to 10 inches long.

Flowers: Several, in loose dangling clusters 2 to 3 inches wide; sepals 5; petals 5, creamy yellow; stamens numerous; stigma 5-toothed; floral bract smooth, dangling, 4 to 5 inches long. May through June.

Fruits: Gray to rusty brown, pea-size, dangling in loose clusters, fused to oblong bract; maturing in October and persisting into winter.

Distribution: Moist wooded slopes, woodland edges, low woods. Uncommon.

In Kentucky: AP, IP.

The sweet fragrance of the flowers have been compared to that of wild grape, honeysuckle, and wild rose and are a favorite honey plant for bees. Basswood nectar is white and said to have a strong flavor and to be of the highest quality. Because of the short flowering time of two to three weeks, the abundance of honey varies from year to year.

Basswood wood is among the lightest and softest of our native trees. Easily worked, it is excellent for carving. The name is a corruption of "bast," referring to the fibrous inner bark.

Other common names include American lime, limetree, and linden. It is often planted as a street or lawn tree.

Common hackberry
Elm Family
Celtis occidentalis L.
Ulmaceae

Key features: Bark warty, often disfigured with clusters of abnormal growth called "witches' broom" in its branches; leaves alternate, short-stalked; fruits reddish black.

Origin: Native.

Life form: A large tree to 80 feet with a rounded crown.

Leaves: Blades ovate, dull green above, smooth to rough, below paler, margins toothed with forward-facing teeth except along the unequal base, tips long-pointed; 2 to 5 inches long.

Flowers: Small, greenish yellow usually with both stamens and pistils, but occasionally with just stamens; appearing at the same time as leaves in spring. April through May.

Fruits: Drupe rounded, fleshy, on stalks to ½ inch long; seed 1, hard, ripening in September and October, often persisting into winter.

Distribution: Moist to dry woods, woodland edges, waterways, roadsides, especially in limestone regions. Common.

In Kentucky: AP, IP, ME.

The genus name, *Celtis,* comes from the Greek name for a tree with sweet fruit. Common hackberry fruits are said to taste like dates and are eaten by birds and other small mammals.

It has very little landscape value today, but at one time the United States Forest Service recommended planting this tree in the prairie states because of its ability to withstand harsh conditions.

The wood is marketed and used like that of elms: for plywood, steam-bent furniture, and veneer.

Winged elm

Elm Family
Ulmus alata Michx.
Ulmaceae

Key features: Small to medium-size tree, with corky wings along 2 sides of the branches; leaves alternate, elliptical to ovate, margins doubly toothed; fruit a samara.

Origin: Native.

Life form: Tree to 50 feet tall.

Leaves: Blades short-stalked, with upper surface dark green, smooth or rough, veins whitish below, bases unequal to rounded, tips pointed; 1 to 3 inches long.

Flowers: Green, inconspicuous, borne in few-flowered clusters in spring. March through April.

Fruits: Samaras lance-ovate, apex with 2 incurved tips, long-hairy on margins; maturing in spring.

Distribution: Dry to moist woods, woodland edges, roadsides. Common.

In Kentucky: (AP-southern section), (IP-absent from northern section), ME.

This family, whose members are trees and shrubs, is economically important because of the superior timber that is produced. The wood is resistant to decay and is used in water-logged conditions and for underwater piles.

Cork elm is another common name and refers to the distinct corky wings on the branches that usually develop in the second year of growth.

Red elm

Elm Family
Ulmus rubra Muhl.
Ulmaceae

Key features: Inner bark mucilaginous; buds dark red, twigs gray; leaves sand-papery rough, hairy below with white tufts of hairs in axils of veins; fruit a samara with smooth margins.

Origin: Native.

Life form: Medium-size tree to 80 feet tall with a broad, open crown.

Leaves: Alternate, short-stalked; blades lanceolate to obovate, margins coarsely toothed, bases unequal, tips abruptly short-pointed; to 6 inches long.

Flowers: Male and female flowers on the same tree in separate clusters of 3 to 5, stalks short to sessile: male with dark red anthers and female with red purple stigmas; appearing before the leaves. March through April.

Fruits: Circular, slightly veined, wings yellow to cream-colored, to ½ inch wide; ripening before or with the unfolding leaves.

Distribution: Moist low woods, upland woods, woodland edges, ravine slopes, waterways, roadsides. Common.

In Kentucky: AP, IP, ME.

Similar species: American elm (*Ulmus americana* L.) is a large, spreading vase-shaped native tree with **reddish brown buds and twigs.** The leaf blades are oblong to ovate, **smooth to slightly rough on upper surface,** dark green and shiny above, paler and hairy below with coarsely toothed margins and unequal leaf bases. The creamy yellow **samaras,** less than ½ inch wide, have **densely hairy margins** and two clawlike wingtips that curve inward; these ripen as the leaves unfold. Cher, Sen, Iroq, Shaw, Chick. Low moist woods, woodland edges, waterways, roadsides, fields, bottomlands. Common. In Kentucky: AP, IP, ME.

The inner bark of red elm is slimy—hence the other common name, slippery elm. This inner bark, which is about ¼ inch thick, softens and dissolves into a gluey slime when chewed or put in water. Native Americans used this slimy inner bark to dress wounds, as it had a cooling effect and helped to prevent inflammation.

Kentucky coffeetree

Legume Family
Gymnocladus dioicus (L.) K. Koch.
Fabaceae

Key features: Tree with scalelike bark peeling outward on the sides; leaves alternate, compound, twice divided, 1 to 3 feet long; fruit a flat, reddish brown pod.

Origin: Native.

Life form: A medium-size tree to 75 feet.

Leaves: Leaflets 40 or more, blades ovate to elliptic, dark blue-green above, margins entire, bases rounded to wedge-shaped, occasionally unequal, tips pointed; 1 to 2 inches long.

Flowers: Male and female on separate trees: male in short clusters, and female whitish, small, fragrant, in panicles 8 to 12 inches long. May through June.

Fruits: Pod broad, leathery, 4 to 8 inches long and 1 to 2 inches wide; seeds few, dark brown, rounded, embedded in a sticky pulp.

Distribution: Low moist woods, upland woods, ravine slopes. Rare at Iroquois; uncommon elsewhere.

In Kentucky: (AP-rare), IP, (ME-rare).

Also called coffeenut and coffeebean, the tree produces large leathery pods that ripen in fall and hang on throughout winter. The seeds somewhat resemble coffee beans. They were used by the first settlers and explorers in Kentucky as a coffee substitute. Seeds are produced in abundance every other year or on a three-year cycle.

The leaves and seeds are poisonous and contain the alkaloid cytisine. Roasting the seeds destroys this toxic property.

Gymnocladus means "naked branch" and describes the stout branches when they are devoid of foliage.

Kentucky coffeetree was the official state "Heritage Tree" for several years.

Black locust

Legume Family
Robinia pseudoacacia L.
Fabaceae

Key features: Medium-size tree with deeply furrowed bark; leaves alternate, leaflets 7 to 19; flowers white, fragrant, in drooping clusters; pods bean-like, brown.

Origin: Native.

Life form: Tree to 75 feet with crooked trunks.

Leaves: Leaflets each 1 to 2½ inches long; blades oblong to elliptic, short-stalked, bluish green above, paler below, margins entire, bases slightly tapered to rounded, tips rounded.

Flowers: Many, in drooping clusters, 4 to 7 inches long; sepals 5, light brown; petals 5, upper rounded petal largest with yellow spot, side and lower ones slightly smaller; stamens 10. April through June.

Fruits: Pods flattened, papery, 2-valved, 2 to 6 inches long; seeds 4 to 8, small, dark brown and mottled.

Distribution: Roadsides, moist woods, thickets, successional lands. Common.

In Kentucky: AP, IP, ME.

This tree is common throughout the state, and when in flower the showy, white, pendulous clusters fill the air with their sweet fragrance. However, in late summer, the leaves are often brown from the *locust leaf miner,* a beetle that feeds on it. These diseased trees stand out, especially along roadsides, because they appear dead.

In Europe, this species is a popular garden ornamental and highly regarded for its timber value. The wood is very hard and strong and used for railroad ties, fence posts, and in shipbuilding.

Bitternut hickory
Walnut Family
Carya cordiformis (Wangenh.) K. Koch
Juglandaceae

Key features: Tall deciduous tree with distinct sulfur yellow, glandular dotted bud scales; leaflets 7 to 9, short-stalked to sessile; fruits nearly round, husk thin.

Origin: Native.

Life form: A tall tree to 75 feet with an irregular, cylindrical crown.

Leaves: Blades ovate-lanceolate, upper surface bright green, smooth, paler below and hairy, margins finely to coarsely toothed, midvein curved, tips long pointed; 3 to 9 inches long.

Flowers: Male and female on same tree; male green, in 3 to 4 catkins hanging from same stalk below the leaves and female inconspicuous, yellow, woolly, above the leaves. April through May.

Fruits: With short tip, husk thin, splitting into 4-winged sections about to the middle; kernel inside reddish brown, bitter tasting.

Distribution: Moist woods, woodland edges. Common at Cherokee; uncommon elsewhere.

In Kentucky: AP, IP, ME.

This beautiful hickory has distinct yellow sulfur buds, although sometimes they are a dull brown due to lack of glandular production.

The Iroquois Indians used the nutmeat oil alone for healthy hair or mixed it with bear grease. The same mix was also used to ward off mosquitoes. The crushed nuts mixed with cornmeal, beans, or berries made a tasty bread or corn pudding. Alone, the nuts are very bitter in taste.

In 1941, Mabel Slack wrote in her master's thesis on the flora of Cherokee Park that hickories were scarce because the hard wood was used for making axe handles, beams of plows, and looms and that this species was the most common.

Pignut hickory
Walnut Family
Carya glabra L.
Juglandaceae

Key features: Large tree with smooth gray bark when young and platelike scales when older; leaves alternate, leaflets 5 to 7; husk woody, pear-shaped.

Origin: Native.

Life form: Tree to 90 feet with an oval crown, the lowermost branches often drooping toward the ground.

Leaves: Terminal leaflet largest, blades dark yellowish green and smooth above, pale below with tufts of hairs in vein axils, margins with sharp incurved teeth; 8 to 12 inches long.

Flowers: Male and female flowers on same tree: male catkins in groups of 3 drooping below the leaves and female flowers, 2 to 10, in clusters or short spikes above the leaves. April through May.

Fruits: Husk 4-valved, brown, usually not splitting to base.

Distribution: Moist woods, uplands, dry ridges. Common.

In Kentucky: AP, IP, ME.

This beautiful hickory is also called broom hickory, swamp hickory, and smoothbark hickory. The heavy, flexible wood was used in the West for making the wheel hubs for covered wagons because of its low conductivity to heat and friction. The Cherokee Indians had several uses for this common tree. The wood ash and water were mixed and used as lye, and the dried nuts and shells were ground into a powdery meal that was used to make soup. A tasty cure for pork was to mix wood ash, salt, and pepper.

Shagbark hickory
Walnut Family
Carya ovata (Mill.) K. Koch.
Juglandaceae

Key features: Bark shaggy when mature; leaflets usually 5, margins finely toothed with tufts of hairs on one or both sides of tooth; fruit a woody husk, 4-ribbed and splitting all the way.

Origin: Native.

Life form: Medium to large tree to 80 feet with distinctive bark separating into long, overlapping loose strips when mature.

Leaves: Alternate, odd-pinnate, 8 to 14 inches long; leaflets mostly 5, the 3 upper leaflets longer than the lower, blades ovate to obovate, dark yellow green above, pale to downy below, bases tapering, tips pointed.

Flowers: Male and female on the same tree opening with or just before the leaves: male in axillary hanging catkins 4 to 5 inches long and female small in sparsely flowered spikes. April through May.

Fruits: Solitary or in pairs; husk thick, yellow brown to dark brown, nearly round, 1 to 2½ inches wide; nut light-colored, ridged on 4 sides.

Distribution: Moist woods, ravine slopes, dry uplands, woodland edges. Common.

In Kentucky: AP, IP, ME.

The distinct shaggy, loose strips of bark are the earmark characteristic for this species. Also called shellbark, scaly-barked, or upland hickory, the green wood is used to impart a unique flavor to "hickory-smoked" hams and bacons.

Native Americans would collect the nuts, pound them into small pieces, strain them, and then preserve the oily parts of the liquid, which they called "hiccory milk." This milk produced a fresh, sweet, rich cream that was a delicacy. The edible nuts were also a popular ingredient for making cakes, candies, and breads.

Black walnut
Walnut Family
Juglans nigra L.
Juglandaceae

Key features: Large deciduous tree with brown furrowed bark; leaves alternate, pinnately compound, 1 to 2 feet long; nut sculptured with ridges, husk light green, fleshy.

Origin: Native.

Life form: Tree to 75 feet or more with round-topped crown and light-colored pith.

Leaves: Leaflets 13 to 23, terminal leaflet often missing or rudimentary; blades ovate-lanceolate, yellowish green above, veins pale below, with glandular hairs, margins toothed, bases rounded or unequal, tips tapering; aromatic when rubbed or crushed.

Flowers: Male and female flowers on the same tree: male catkins separate, dangling, 3 to 5 inches long and female flowers solitary or several in a cluster at the tip of the branches. April through May.

Fruits: Nut rounded, $1\frac{1}{2}$ to 2 inches wide, surrounded by a husk that ripens in fall and drops, turning brown with age.

Distribution: Upland woods, ravine slopes, along waterways. Common.

In Kentucky: AP, IP, ME.

This valuable tree is known for its dark brown, strong, close-grained wood and is used for clocks, veneer, gun stocks, piano cases, and fine furniture and cabinets.

The Iroquois Indians used the bark as a medicine for toothaches and the boiled and charred young twigs and old bark were steeped in water and applied to snakebites.

The leaves contain juglone, a sedative comparable with diazepam or Valium based on recent studies with animals. Traditionally, Native Americans made a poultice from the bark and applied it for "craziness."

A tea made from the leaves was used for its insecticidal properties to keep away bedbugs.

Tree-of-heaven

Quassia Family
Ailanthus altissima (Mill) Swingle
Simaroubaceae

Key features: Tree with stout, crooked, ill-scented branches; leaves alternate, pinnately compound, 1½ to 3 feet long; flowers in large panicles; fruits winged, twisted.

Origin: Eastern Asia.

Life form: Tree to 75 feet tall with oval crown.

Leaves: Leaflets 11 to 41, short-stalked; blades ovate-lanceolate to oblong, margins entire, bases rounded or unequal with 1 to 3 glandular teeth near the base; each leaflet up to 6 inches long.

Flowers: Yellowish green, male and female on separate trees in large panicles to 20 inches long; male flowers emit a fowl odor. June.

Fruits: Oblong, green at first, turning tan to brownish red in late summer and fall, persisting through most of winter; about 2 inches long.

Distribution: Disturbed ground in open woods, woodland borders, roadsides. Common. Trees are being cut down due to their invasive tendencies.

In Kentucky: (AP-rare), IP, (ME-rare). Listed as a Severe Threat by the Kentucky Exotic Pest Plant Council.

The seeds of this tree from China were first brought over to Europe by the great naturalist Peter Collins in 1751 and to North America in 1784. Nurseries in larger cities along the East Coast began selling this hardy tree because of its adaptability, fast growth, resistance to pests, and tolerance to pollution. In the 1850s, it was brought into the western United States by Chinese immigrants during the Gold Rush, possibly for medicinal uses. Since then, it has escaped and naturalized throughout the country. Once established, it is difficult to eradicate because of the root sprouts which can form dense colonies. It is also a heavy seed producer.

Ailanthus contains quassinoids, chemicals that can cause heart problems, severe headaches, and nausea in people who do not protect themselves from the sap.

Silver maple

Maple Family
Acer saccharium L.
Aceraceae

Key features: Large tree with smooth gray bark when young becoming flaky when older; leaves opposite, deeply 5-lobed, silvery white below; fruit a winged samara.

Origin: Native.

Life form: Medium to large tree to 90 feet with slender branches that sweep downward then curve upward at the tips.

Leaves: Blade deeply 5-lobed, with middle lobe flaring above middle, narrower below, sinuses U-shaped, lobes coarsely toothed, tips sharp-pointed, bright green shiny above, leaf stalks drooping; 3 to 7 inches long.

Flowers: Greenish yellow, in clusters, either all male or all female on the same tree. February to April.

Fruits: Samara pale brown, wings spreading, 1½ to 2½ inches long; falling before leaves are fully expanded.

Habitat: Low moist woods, bottom-lands, waterways, fields, often escaping from cultivation. Common.

In Kentucky: AP, IP, ME.

This species is often planted as a street tree because of its fast growth, but the weak long and slender branches are easily damaged by wind and ice storms—hence the name soft maple. It also has a distinctive pinkish color to the inner bark and a ring of darker, curled-up-looking bark around the base that help to identify this species.

The Iroquois Indians used the dried bark and pounded it into a flour to make bread.

Sugar maple
Maple Family
Acer saccharum Marshall
Aceraceae

Key features: Tree with smooth gray bark when young, turning black with scaly ridges curling outward from the sides when mature; leaves opposite, 3-to-5-lobed, sinuses U-shaped, buds sharp.

Origin: Native.

Life form: Large tree up to 100 feet tall with a broadly rounded crown when grown in the open, more upright in woods but maintaining a deep crown.

Leaves: Blades smooth, palmately 3-to-5-lobed with 1 main vein in each lobe originating at the leaf base, sinuses U-shaped, tips pointed, dark green above, paler beneath, margins few-toothed; 3 to 5 inches long and wide.

Flowers: May be male, female, or bisexual in separate dangling clusters on same tree; sepals and petals absent; stamens greenish yellow; pistils greenish yellow. April through May.

Fruits: Samaras 1 to 1½ inches long, with widely spreading, curved wings; seed plump.

Distribution: Moist to dry woods, fields. Common to abundant, particularly in deep, well-watered, and nutrient-rich soils. In some areas, seedlings form a monoculture in the understory because they are able to survive in very low-light conditions.

In Kentucky: AP, IP, ME.

Similar species: **Red maple** (*Acer rubrum* L.) is a **medium to large tree to 70 feet tall with smooth, light gray bark when young becoming darker gray and flaky when mature.** The **regularly toothed leaves are palmately 5-lobed** and **smooth** (var. *rubrum*) **or 3-lobed and hairy** (var. *trilobum*). Leaf stalks are usually red and the **sinuses V-shaped. The leaf buds are blunt.** Leaves turn brilliant red in autumn. Samaras less than 1½ inch long. Cher, Sen, Iroq, Shaw, Chick. Moist to wet woods, along waterways, swamps. Uncommon. In Kentucky: AP, (IP-rare in Inner Bluegrass), ME.

The sugar maple is a familiar sight throughout the eastern half of the United States. In autumn, the yellow to orange leaves add brilliant color to the landscape.

Native Americans and early settlers learned how to make syrup and sugar from the sap, as well as a beer. The syrup was used as a liver tonic and kidney cleanser. The Iroquois Indians made a compound infusion from the bark and used the drops for blindness and the sap for sore eyes.

Flowering dogwood

Dogwood Family
Cornus florida L.
Cornaceae

Key features: Understory tree with checkered bark; leaves opposite, parallel-veined; flowers small, surrounded by four white petal-like bracts.

Origin: Native.

Life form: Small to medium-size tree to 40 feet tall with spreading crown.

Leaves: Blades oval to ovate, veins distinct with 5 to 6 pairs curved toward apex, dark green above, paler below, margins entire, bases rounded to wedge-shaped; 3 to 5 inches long.

Flowers: Small, greenish yellow in rounded clusters surrounded by showy petal-like bracts. April through May.

Fruits: In dense clusters of 3 to 5, each ovoid, tipped with a persistent style; turning scarlet red when mature.

Distribution: Moist to dry woods, woodland edges, fields. Common at Iroquois; uncommon elsewhere.

In Kentucky: AP, IP, ME.

Considered one of the most popular and beautiful of all our native trees because of the showy white bracts and scarlet red fruit, this once common species has become uncommon or rare due to the deadly fungus called "dogwood anthracnose" and "dogwood powdery mildew." Although some native populations appear to be resistant, damage is apparent in the forests of the eastern United States. Cultivars that are disease resistant are plentiful and available throughout the nursery trade.

In 1939 and 1940, Frederick Law Olmsted's firm planted about 500 dogwoods in Cherokee.

European buckthorn

Buckthorn Family
Rhamnus cathartica L.
Rhamnaceae

Key features: Large shrub, branches thorny and tipped with a stout spine; leaves opposite to subopposite, ovate to elliptic, dull green above; fruits round, black.

Origin: Europe.

Life form: Deciduous, irregularly branched, large shrub to small tree to 20 feet; branches crooked, often thorny.

Leaves: Short-stalked; blades dull green above, paler below, margins toothed, veins in 3 to 5 pairs, bases rounded to wedged shaped, tips short-pointed to round; 1 to 3½ inches long.

Flowers: Small, greenish yellow, few-flowered, in axillary clusters, male and female on separate plants. May through June.

Fruits: Round black drupe berry-like, ¼ inch wide; seeds 3 to 4 plump, top with deep groove.

Distribution: Open disturbed woods, woodland edges, thickets, roadsides. Common, but being eradicated due to its invasive qualities.

In Kentucky: (IP-mostly Bluegrass). Listed as a Severe Threat by the Kentucky Exotic Pest Plant Council.

Similar species: Dahurian buckthorn (*Rhamnus davurica* Pall.) is native to northeastern Asia and a **large shrub or small tree** growing to 30 feet tall. The leaf **blades,** 2 to 6 inches long, are **obovate to elliptical, glossy bright green above and gray-green below** with finely toothed margins. It is naturalized in the same areas in Kentucky as European buckthorn and often grows in the same habitat. At Cherokee, both species are found growing together in open disturbed woods along roadsides, although Dahurian buckthorn is uncommon. In Kentucky: (IP-mostly Bluegrass; rare). May through June. [*syn=Rhamnus citrifolia* (Weston) W. J. Hess & Stern]

European buckthorn is a medicinal plant and was cultivated for centuries in Europe and Asia. The flowers and very bitter-tasting fruits were used as a strong laxative; it also provided a natural yellow dye. It was introduced into the United States by the early settlers as a hardy hedge plant and has now escaped and become naturalized in disturbed habitats in the Bluegrass.

Boxelder maple

Maple Family
Acer negundo L.
Aceraceae

Key features: Tree with distinct new green growth in spring; leaves opposite, compound, leaflets 3 to 7; fruits winged, dangling from slender stalks.

Origin: Native.

Life form: Small to medium-size tree to 60 feet.

Leaves: Short-stalked, blades ovate, oblong or elliptic, margins coarsely toothed to slightly lobed; 2 to 4 inches long.

Flowers: Male and female flowers usually on separate trees, but on some trees some flowers are bisexual, all opening before the leaves, male flowers in large umbel-like clusters and female flowers in dangling clusters on slender stalks. April through May.

Fruits: Wings slightly spreading to a 45-degree angle or less, green at first, turning light brown when mature; 1 to 2 inches long.

Distribution: Moist to wet open woods, swamps, along waterways. Often a weedy invader of disturbed sites. Common.

In Kentucky: AP, IP, ME.

Boxelder maple is an opportunistic species and the mostly widely distributed maple in North America. It was introduced into England in 1688 and grown at Bishop Compton's garden at Fulham as an ornamental. It doesn't resemble a "typical" maple when in leaf, but when the fruits mature it is clear to see.

The Native Americans burned the wood for incense and for making spiritual medicines, and the charcoal was used for ceremonial painting and tattooing.

White ash
Olive Family
Fraxinus americana L.
Oleaceae
(*syn=Fraxinus biltomoreana* Beadle)

Key features: Tree with silvery gray bark with diamond-shaped furrows; leaves opposite, compound, to 15 inches long; fruit with a single wing.

Origin: Native.

Life form: A large tree to 100 feet or more with straight limb-free trunks.

Leaves: Leaflets 5 to 9, blades ovate, dark green above, whitened below, margins entire to slightly toothed, bases rounded or tapered to a short stalk, tips pointed; 8 to 12 inches long.

Flowers: Male and female in dense clusters usually on separate trees, opening before the leaves; male with stamens, anthers linear and female with single style, stigma 2-cleft. March through April.

Fruits: Tannish brown wing attached to the bottom of swollen seed; 1 to 2 inches long.

Distribution: Low moist woods, dry to moist uplands. Common.

In Kentucky: AP, IP, ME.

A valuable timber tree, the wood is strong and hard and used for making baseball bats, furniture, railroad ties, crates, and many other products. The Native Americans used the wood for canoes and snowshoes.

In Appalachia, the chewed bark was used to heal sores, and a tea was made from the buds to treat snakebites.

Bladdernut
Bladdernut Family
Staphylea trifolia L.
Staphyleaceae

Key features: Shrub with striped bark; leaflets 3; flowers bell-shaped, white, in few-flowered, drooping clusters; fruit capsules inflated, balloon-like.

Origin: Native.

Life form: Shrub, often multi-stemmed, loose and open, to 15 feet tall with suckering root-shoots.

Leaves: Leaflets 3, terminal leaflet on longer stalk then 2 laterals, blades ovate, lanceolate to elliptic, dark green above, lighter green with white fuzz below, margins finely toothed, bases unequal; 2 to 4 inches long.

Flowers: Sepals 5, green with white margin; petals 5, tip rounded or pointed; stamens 5, alternate with the petals; pistil 1, styles 3. April through May.

Fruits: Capsule 3-lobed, pale green changing to brown, about 2 inches long; seeds several, rounded, hard, glossy brown.

Distribution: Moist wooded slopes, creek banks, rocky woods. Uncommon.

In Kentucky: AP, IP, ME.

This plant family includes trees and shrubs found in temperate and tropical regions of the world. The name *Staphylea* comes from the Greek word *staphyle* "a cluster" and describes the arrangement of the flowers. The common name bladdernut refers to the inflated fruits.

Ohio buckeye

Buckeye Family
Aesculus glabra Willd.
Hippocastanaceae

Key features: Leaves opposite, leaflets 5; flowers greenish yellow, in erect terminal clusters 4 to 7 inches long; fruits light brown, prickly.

Origin: Native.

Life form: A small to medium-size tree to 60 feet tall with ashy gray bark.

Leaves: Blades elliptic to oval, upper surface dark green, lower hairy when young, becoming smooth, margins finely toothed, bases narrowed, tips tapered; 3 to 6 inches long.

Flowers: Sepals 5, tubular; petals 4, greenish yellow, unequal; stamens 6 to 8. April through June.

Fruits: Capsule roundish, leathery; single seed large, brown, smooth, shiny; about 2 inches long.

Distribution: Waterways, dry to moist wood, in calcareous soils. An especially beautiful stand is found in Seneca, where typical tree height is 50 feet; absent from Iroquois.

In Kentucky: (AP-rare), (IP-mostly Bluegrass), (ME-rare).

Ohio buckeye is the first tree to leaf out in the woods in late March and the first to lose its leaves in fall. It is also called fetid or stinking buckeye because stems have a skunk-like odor when broken. The wood is light, soft, and weak and is used in making cheap furniture, boxes, and crates.

The seeds are poisonous and contain saponins. The young shoots in spring and the seeds are poisonous to both humans and livestock. The symptoms include vomiting, depression, loss of coordination, twitching of muscles, and paralysis.

Native Americans ground up the seeds and used the powder to kill fish, which is a common use of saponin-rich plants.

Carex

Sedge Family
Carex albursina Sheld.
Cyperaceae

Key features: Stems tufted; leaves lax, light green; flowers inconspicuous; fruits inflated, beak abruptly bent.

Origin: Native.

Life form: Perennial herb from fibrous roots.

Stems: Lax, winged, basal sheath papery, brown above, reddish purple below.

Leaves: Blades flat, ascending to lax, midrib distinct with 2 lateral veins; stem leaves alternate, shorter; to 2 inches wide.

Flowers: In terminal and lateral spikes with broad leaflike bracts exceeding the spikes, often concealing them; inflated fruits 3 to 20 per spike, green, obovate, overlapping, beaked. April through June.

Fruits: Achene ovoid.

Distribution: Moist wooded slopes, especially in limestone soils. Uncommon.

In Kentucky: AP, IP, ME.

The *Cyperaceae* is a large family with species found in all parts of the world, especially in damp or wet regions of the temperate and subarctic zones. It is also an economically important family with members used for a variety of purposes in providing food, fiber, medicine, and forage. Some species are cultivated for water garden ornamentals.

This woodland sedge with wide leaves is attractive and an exciting find in Cherokee and Iroquois.

Straw-colored flat-sedge

Sedge Family
Cyperus strigosus L.
Cyperaceae

Key features: Plant from rhizomes; stems triangular, light green; leaves grasslike, 3-ranked; flower clusters of straw-colored to pale brown spikelets.

Origin: Native.

Life form: Tufted perennial.

Stems: Erect, often equal to the basal leaves, from an enlarged, reddish purple corm-like base; 10 to 30 inches tall.

Leaves: Blades flat, firm, linear, margins entire, surface rough, equaling or surpassing the inflorescence.

Flowers: In loose, oblong-cylindric clusters on stalks to 5 inches long; spikelets mostly horizontally radiating or upper ascending, lower descending. July through September.

Fruits: Achenes 3-angled, purplish brown, narrowly oblong, surface pitted.

Distribution: Disturbed moist ground, cultivated beds, turf, waterways, roadside ditches, fields, pond margins. Common.

In Kentucky: AP, IP, ME.

Similar species: Chufa (*Cyperus esculentus* L.) is a perennial herb bearing **small, dark underground tubers on the stolons.** The yellowish green triangular stems have grasslike leaves that are mostly basal. The spikes are ovoid to hemispheric and made up of **clusters of yellowish brown to dark brown,** linear, very small spikelets that emerge at right angles to the stalk. Disturbed ground as above. Common weed in all parks. July through September. In Kentucky: AP, IP, ME.

Chufa is considered to be one of the world's worst weeds in agricultural lands. Also known as yellow nutsedge, northern nutsedge, earth almond, and tigernut, it is an ancient crop of the Middle East and Egypt, found in tombs 4,000 years old. In southern Europe, western Asia, and parts of Africa, it is grown for its edible tubers, which are roasted and eaten like potatoes but have a nutlike taste. They may be cooked, ground into a flour, or made into a cold drink. The species name means "edible."

The genus *Cyperus* is large and found throughout the world. The species are difficult to identify because of the small, inconspicuous flowers.

Path rush

Rush Family
Juncus tenuis Willd.
Juncaceae

Key features: Plants tufted, often forming dense colonies; stems wiry, of unequal lengths; flowers inconspicuous, green to brown.

Origin: Native.

Life form: Perennial herb from fibrous roots and rhizomes.

Stems: Erect or ascending, rounded, smooth; 4 to 22 inches tall.

Leaves: Basal, blades flat to very narrow, smooth, ascending, about half as long as stems; membranous ear-shaped lobes at leaf base longer than sheath apex.

Flowers: Few, in terminal clusters subtended by 2 to several leaflike bracts that surpass the flowers; tepals 6, the outer 3 white-scaly with green stripe, inner 3 similar but with long-pointed tips; stamens 6; stigmas 3. June through September.

Fruits: Capsule rounded, straw-colored; seeds tiny, dark, asymmetrical, veined, beaked.

Distribution: Disturbed compacted ground in dry to wet soils, especially along paths, thin soil under trees. Common.

In Kentucky: AP, IP, ME.

This weedy species belongs to a family that occurs worldwide chiefly in wet or damp habitats. It has spread from North America to other continents, including Eurasia and Australia. Other common names include wire-grass, poverty rush, and field rush.

Some rushes are woven into mats, ropes, baskets, and chair seats, and the pith has been used for candlewicks. The genus name, *Juncus*, is from the Latin word *jungere*, "to join" and refers to their use in rope making.

Rushes are difficult to identify. They have small, inconspicuous flowers with a dried, scaly appearance. It is important to look at the apex of the basal sheath for proper identification.

Prickly woodrush

Rush Family
Luzula echinata (Small) F. J. Herm.
Juncaceae

Key features: Loosely tufted perennial; leaves basal, grasslike with long wispy white hairs; flowers inconspicuous, green turning brownish tan, in loose clusters.

Origin: Native.

Life form: Perennial herb from a rhizome.

Leaves: Blades narrow, flat, tapering at tip, margins entire with long white hairs; stem leaves few, similar but smaller; to 6 inches long.

Flowers: Tiny, in 2 to 16 spikes on threadlike simple or forking rays that are ascending, horizontal, or reflexed; central spike in cluster stemless or nearly so; tepals 6, margins clear; involucral bracts, tiny, papery; $1/16$ wide. April through June.

Fruits: Capsules round, usually shorter than tepals; seeds tiny, brown.

Distribution: Open wooded slopes, along paths, base of trees in shady turf. Common.

In Kentucky: AP, IP, ME.

This plant family occurs mostly in moist, cold temperate or montane regions worldwide, and most species have small, inconspicuous flowers that are easily overlooked. The genus *Luzula* is from *Gramen Luzulae* or *Luxulae,* diminutive of *lux* "light," and refers to one of the species that appeared shining in the morning dew. (The long white hairs viewed in the morning light truly shine.)

Origin: Native.

Life form: Perennial grass from fibrous roots and rhizomes.

Stems: Erect, stout, solid, often bluish purple with white tinge; 3 to 6 feet tall.

Sheaths: Smooth to slightly hairy.

Leaves: Blades linear, floppy, silky-hairy at base; to 2 feet long.

Ligules: Membranous, fringed, very small.

Spikelets: In pairs, 1 sessile, 1 stalked, dull greenish gray to purplish red, awn bent, twisted below. June through September.

Fruits: Grain dull brown, long-pointed, longitudinally grooved.

Distribution: Abundant prairie grass at Summit Field.

In Kentucky: (AP-rare), (IP-rare in inner Bluegrass), ME.

Similar species: Broom-sedge (*Andropogon virginicus* L.) is a native tufted warm season perennial grass turning orange-brown when dry. The **inflorescence is made up of short clusters scattered among leaves along the stem,** each cluster with conspicuous cottony hairs. The spikelets are very narrow, straw-colored, pointed with a protruding bristle. Cher. Iroq. Fields, roadsides. Common. August through November. In Kentucky: AP, IP, ME.

Big bluestem

Grass Family
Andropogon gerardii Vitman
Poaceae

Key features: Warm season grass often forming large clumps; leaves alternate, narrow, dull green to bluish; inflorescence made up of a few V-shaped racemes at the top of the stalk.

This impressive grass is rightly called big bluestem because of the size of the plant and the color of the leaves and stem. Another common name is turkey-foot grass, which describes the seed head that usually branches into three parts resembling a turkey's foot. It is a dominant species of the tallgrass prairies of the Midwest and a major forage and hay grass throughout that region. Few prairie grasses can equal big bluestem in quality and quantity of forage produced, and it is relished by livestock, bison, and other mammals. It is a popular grass for prairie restoration sites.

Giant cane
Grass Family
Arundinaria gigantea (Walter) Muhl.
Poaceae

Key features: Large woody grass; branches forming fanlike clusters; leaves long, narrow.

Origin: Native.

Life form: Perennial grass from tough rhizomes.

Stems: Erect, hollow, round, unbranched at first, branching later; 5 to 6 feet tall, but can grow to 25 feet. (The height is determined in first growing season and no extra growth in height after that).

Sheaths: On stem branches, loose and papery; on leaves, overlapping with bristles on summit.

Leaves: Blades long, lanceolate, smooth above, short-hairy below, midvein white, base rounded to tapering, tip sharp-pointed; to 12 inches long.

Ligule: Short, firm, white-papery.

Spikelets: Eight- to 12-flowered on long, slender stalks; rarely flowers, possibly only every 40 years.

Fruits: Grain ellipsoid to round, ½ inch long.

Distribution: Moist wooded slopes. Rare.

In Kentucky: AP, IP, ME.

Giant cane, or bamboo, is an important native grass of moist bottomlands and forest understory. This species once covered thousands of acres of land in Kentucky as well as throughout southeastern North America, forming dense stands called "canebrakes." Early accounts describe these brakes as being so thick and tall that it was difficult to pass through them. These massive, dense thickets were a major component of the understory. Their disappearance has been due in part to overgrazing and conversion of land into agriculture. Sadly, only a few remnants remain today.

Also called southern cane or cane reed, the round hollow stems were used as fishing poles and pole bean supports.

This plant is rare in Cherokee and Seneca, but as part of the restoration efforts, giant cane is being planted along some areas of Beargrass Creek.

River wood oats

Grass Family
Chasmanthium latifolium
(Michx.) H.O. Yates
Poaceae
(*syn=Uniola latifolia* Michx.)

Key features: A grass, often forming colonies; inflorescence an open nodding panicle bearing a few flat, oval spikelets.

Origin: Native.

Life form: Perennial grass from rhizomes.

Stems: Single or little branched, smooth, light green, leafy; 1 to 3½ feet tall.

Sheaths: Smooth.

Leaves: Alternate, blades linear-lanceolate, flat, light green, mostly smooth or with a few hairs at base of leaf blade; up to 10 inches long.

Ligule: Very small, membranous.

Spikelets: Compressed, 6- to 17-flowered, broad, flat, nodding, light green at first turning brownish tan at maturity, about 1 inch long. July through October.

Fruits: Grain brown, small.

Distribution: Shady moist woods, waterways, pond margins. Locally common.

In Kentucky: AP, IP, ME.

This attractive clump-forming grass grows along stream and river banks in rich woods throughout the southeastern and south-central parts of the United States. There are five species in the genus, and several are sold in the nursery trade for ornamental plantings.

Other common names include spangle grass, and wild oats, river oats, and flathead oats, referring to the similarity of the "flowers" to that of cultivated oats.

Bermuda grass

Grass Family
Cynodon dactylon (L.) Pers.
Poaceae

Key features: Warm-season grass from rhizomes and stolons, often mat-forming; inflorescence made up of 3 to 7 fingerlike clusters that radiate from the end of the stem.

Origin: Eurasia.

Life form: Perennial grass.

Stems: Erect or ascending, often rooting at the nodes, wiry; to 12 inches tall.

Sheaths: Smooth except with a few hairs at the apex and margin.

Leaves: Blades linear-lanceolate, gray-green, hairy or smooth except for a small tuft of hairs just above the collar; 1 to 4 inches long.

Ligules: A ring of very short white hairs.

Spikelets: Straw-colored to brown, flattened, pointed, and crowded on one side of each branch of the cluster. July through September.

Fruits: Grain, oval, reddish orange, tiny.

Distribution: Disturbed ground, especially troublesome in turf, cultivated beds. Abundant.

In Kentucky: AP, IP, ME.

Bermuda grass which is often planted in lawns and golf courses, is also used as a pasture and forage grass and considered to be one of the worst hay fever–causing grasses in the United States. Exactly when it arrived in America is uncertain. James Meese, in his "Geological Account of the United States" (1807), stated, "Probably as important a grass as any in the southern states is Bermuda Grass, which grows with great luxuriance and propagates with astonishing rapidity by means of its numerous jointing, every one of which takes root."

This species is considered a weed in many countries and is very difficult to eradicate once it is established.

A good identification characteristic is that the creeping stolons have a dead, tan, bladeless sheath at each joint.

Orchard grass

Grass Family
Dactylis glomerata L.
Poaceae

Key features: Cool-season grass; leaves light green to dark blue-green; spikelets fan-shaped, in dense, 1-sided clusters on a stiff panicle branch.

Origin: Europe.

Life form: Tufted, perennial grass from fibrous roots.

Stems: Erect, light green, round, smooth or rough; 2 to 4 feet tall.

Sheaths: Dull green, longitudinally veined, flattened, rough.

Leaves: Basal, in dense clumps, stem leaves alternate, few; blades flat or folded, midrib rough, distinct, slightly rough on upper and lower surface; 6 to 12 inches long.

Ligules: Membranous, rounded to pointed, becoming shredded.

Spikelets: Few, crowded, greenish white, turning brown at maturity on panicle branches 4 to 10 inches long. May through June and sporadically until September.

Fruits: Grain elongated, tip bent off center, short-awned.

Distribution: Disturbed ground. Common.

In Kentucky: AP, IP, ME.

This European clump-forming grass spreads by tillers and has been cultivated in North America since 1760. It was first introduced into Virginia and used for pasture, hay, and silage. It is a common weed throughout most of the United States and grows in light shade to open habits, such as orchards—hence the name orchard grass. *Glomerata* means "gathered in bunches" and refers to the 1-sided flower cluster. Another common name is cock's foot, which describes the shape of the flower head with long lower branches that stick out to the side.

Life form: Summer annual from fibrous roots.

Stems: Simple or several, often bent at the base (like elbows), green to reddish purple; 2 to 4 feet tall.

Sheaths: Smooth, often with small glands above.

Leaves: Smooth surface, narrow, margins sharp, midvein prominent, tapered to a point; 8 to 16 inches long.

Ligules: None.

Spikelets: Many, overlapping, oval, densely arranged on one side of upright or nodding branches, 4 to 8 inches long. June through October.

Fruits: Grain tan, shiny, longitudinally ridged on rounded side.

Distribution: Moist to wet disturbed ground, waterways, pond margins, roadside ditches, low moist woods, fields. Common weed.

In Kentucky: AP, IP, ME. Listed as a Significant Threat by the Kentucky Exotic Pest Plant Council.

Common barnyard grass

Grass Family
Echinochloa crus-galli (L.) P. Beauv.
Poaceae

Key features: Tufted warm season grass; leaves have no ligules or auricles; spikelets green to purplish with bristles short, long, or absent.

Origin: Eurasia.

Producing over 40,000 seeds per plant, Native Americans used the seeds, which were pounded, winnowed, parched, and ground into a meal, for a food, alone or with fish. This species is related to the economically valuable Japanese millet, which is a staple, high-quality cereal native to the temperate regions of Asia. This species is widely established throughout the United States and grows in moist, disturbed sites.

Bottlebrush grass

Grass Family
Elymus hystrix L.
Poaceae
(*syn=Hystrix patula* Moench)

Key features: A warm-season grass; inflorescence with greenish white spikelets pointing away from stem, looking like a "bottle brush."

Origin: Native.

Life form: Perennial grass with fibrous roots.

Stems: Green, nodes swollen; 2 to 4 feet tall.

Sheaths: Smooth, often with whitish tinge, finely ribbed.

Leaves: Alternate, broad, flat, grayish green, smooth, lax; to 12 inches long.

Ligules: Short, membranous.

Spikelets: Few, widely spaced, horizontally spreading when mature, usually in pairs, 2-to-4-flowered, in an erect or slightly drooping spike 4 to 7 inches long; greenish white, becoming tan at maturity, awns long, needlelike; June through July.

Fruits: Grain long, narrow.

Distribution: Moist woods, open rocky woods, woodland edges. Uncommon.

In Kentucky: AP, IP, ME.

Bottlebrush grass is distinctive looking and easily recognized by the "bottle brush" appearance of the spike. The genus, *Hystrix,* is Greek for "hedgehog" or "porcupine" and describes the stiff, bristly spikes.

Nepalese eulalia

Grass Family
Microstegium vimineum (Trin.) A. Camus
Poaceae
[*syn=Eulalia viminea* (Trin.) Kuntzel]

Key features: Straggling tufted grass forming dense colonies often in moist ground; leaves alternate, asymmetrical with distinct, off-center white midrib; flowering in fall.

Origin: Southeast Asia.

Life form: Annual grass from fibrous roots.

Stems: Light green, wiry, lower portions sprawling, rooting at the nodes, terminal portions ascending to erect; 2 to 3½ feet tall.

Sheaths: Shorter than internodes, hairy at summit.

Leaves: Blades lanceolate, short, flat, tapering at both ends, margins smooth; 1 to 4 inches long.

Ligules: Membranous, short, margins jagged.

Spikelets: In 1 to 3 terminal or axillary clusters, erect, or ascending branches to 3 inches long; fertile spikelets falling off early, 1 floret clasping and 2 stalked. September through October.

Fruits: Grain elliptic, reddish yellow, small.

Distribution: Moist to wet shady disturbed woods, woodland edges, roadside ditches, waterways, floodplains, especially in alluvial soils. Especially troublesome in Cherokee.

In Kentucky: AP, (IP-mostly Bluegrass Region), ME. Listed as a Severe Threat by the Kentucky Exotic Pest Plant Council.

Also called packing grass and Japanese stiltgrass, this Asian grass was first introduced into Tennessee in 1919, where it was often used as cushioning material in packing fine porcelain from China. It has since escaped and spread throughout much of the eastern United States where it often forms dense stands in suitable habitats by rooting at the nodes and producing abundant seed, up

Invasive plant
Cher, Sen, Iroq, Shaw, Chick

to 1,000 per plant. The seeds spread by floodwaters, heavy rains, and soil movement and can remain viable in the soil for five years or more. Although this species prefers shady, wet areas, this opportunistic plant can also be found in sunny, dry locations.

The genus name comes from the Greek *micros* "small" and *stege* "cover," and refers to the small glumes.

Switch grass

Grass Family
Panicum virgatum L.
Poaceae

Key features: Warm-season grass, often clump-forming; leaves alternate, linear, bluish green; inflorescence an open, freely branched panicle.

Origin: Native.

Life form: Perennial grass from vigorous rhizomes.

Stems: Solitary or in dense leaf clumps; 3 to 6 feet tall.

Sheaths: Smooth or long hairy, margins usually short-hairy.

Leaves: Blades flat, erect, ascending or spreading, midrib distinct, bases rounded to slightly narrow; 5 to 16 inches long.

Ligules: Hairs dense, silky, short.

Spikelets: Ovoid, borne at end of thin, straight branches.

Fruits: Grain pink or dull purple to golden brown when mature, very small.

Distribution: Open fields, along waterways. Uncommon at Cherokee; abundant prairie grass at Summit Field.

In Kentucky: AP, IP, ME.

This airy, beautiful grass is one of the dominant species of the central North American tallgrass prairies along with Indian grass, big and little bluestem, side-oats grama, and grama grass. These extensive prairies were plowed under as the European settlers began to move westward. Soon, the land was planted in crops such as corn, wheat, and oats.

Switchgrass is an important forage grass and has been introduced into other parts of the world for its palatable and nutritious qualities.

Yellow foxtail

Grass Family
Setaria pumila (Poir.) Roem. & Schult.
Poaceae
[*syn=Setaria glauca* (L.) P.Beauv.]

Key features: A grass with dense yellowish green cylindrical flower clusters at top of stem; bristles 5 to 20.

Origin: Eurasia.

Life form: Summer annual grass from fibrous roots.

Stems: Erect, ascending to sprawling on ground, branched at base, smooth; 1 to 2½ feet tall.

Sheath: Flattened, keeled along margins, smooth, often reddish at base.

Leaves: Blades flat or folded, narrow, tapering at apex, midrib white to pale green, smooth except for hairs at the leaf base; 3 to 10 inches long.

Ligule: Membrane fringed with very fine hairs.

Spikelets: Clusters erect, 2 to 4 inches long; subtended by 5 to 20 yellow-green to yellow-brown bristles; lemnas transversely wrinkled. June through October.

Fruits: Grain tiny, broadly ovate, slightly glossy, yellow with tiny black markings.

Distribution: Disturbed ground along roadsides, in turf, fields, thickets, pond margins. Common.

In Kentucky: AP, IP, ME.

Similar species: Giant foxtail (*Setaria faberi* Herrm.) is a summer annual grass from China with **long green, cylindrical flower clusters nodding** at the top of the stem. Spikelet subtended by **3 to 6 bristles.** The ligule is a ring of short, white hairs. Cher, Sen, Iroq, Shaw, Chick. Disturbed ground. June through October. In Kentucky: AP, IP, ME. Listed as a Significant Threat by the Kentucky Exotic Pest Plant Council.

There are several weedy species of foxtail in the United States. Giant foxtail, named for its discoverer, Ernest Faber, was introduced into the United States from China, probably through contaminated Chinese millet seed. It was first reported in 1939 in northern Virginia and has since spread throughout most of the eastern United States. In Kentucky it is one of the most troublesome weeds.

Indian grass

Grass Family
Sorghastrum nutans (L.) Nash
Poaceae

Key features: Warm-season grass; stems clump-forming; spikelets produced in a narrow terminal panicle to 15 inches long.

Origin: Native.

Life form: Perennial, tufted grass from fibrous roots and short, scaly rhizomes.

Stems: Erect, usually not branched, smooth, light green with dark, swollen, silky nodes; 3 to 8 feet tall.

Sheaths: Dull green, open, smooth.

Leaves: Mostly at base, alternate on stem; blades bluish green, flat, spreads upward and outward from stem; 4 to 12 inches long.

Ligules: Clawlike.

Spikelets: In short racemes, golden brown, narrow, slightly curved, awns twisted below, bent above about ½ inch long. August until frost.

Fruits: Grain reddish brown, finely ridged, widest at top, tapering at base.

Distribution: Abundant prairie grass at Summit Field; rare at Cherokee.

In Kentucky: (AP-rare), (IP-mostly Outer Bluegrass), ME.

This beautiful native grass is also known as yellow Indian grass and once dominated millions of acres in the prairie region in central United States prior to agriculture. Highly adaptable and capable of enduring drought, it is used as a forage crop, for erosion control on slopes, and in tallgrass prairie restorations. It grows readily by seed in pure stands or mixed with other prairie grasses such as big bluestem, little bluestem, and switch grass.

A spectacular display can be seen in late summer and fall at Summit Field.

Johnson grass

Grass Family
Sorghum halepense (L.) Pers.
Poaceae

Key features: Warm-season grass from thick, creeping rhizomes; leaf has a prominent white midvein; inflorescence open, much-branched.

Origin: Mediterranean Region.

Life form: Perennial grass.

Stems: Stout, erect, and hairy at the nodes; to 8 feet tall.

Sheaths: Smooth, green to purplish, open, often with few hairs on the margin.

Leaves: Blades narrow, flat, smooth; to 23 inches long.

Ligules: White, membranous, finely toothed at top; older ligules have fringe of hairs on top half.

Spikelets: In pairs along the branches of the open, whorled panicle; one of each pair stalkless, the shorter, wider floret produces seed and has a twisted bristle-like awn. June through November.

Fruits: Grain reddish brown, oval, slightly shiny.

Distribution: Disturbed ground in fields, roadsides, thickets, woodland edges, low woods, waterways, cultivated beds.

In Kentucky: AP, IP, ME. Listed as a Severe Threat by the Kentucky Exotic Pest Plant Council.

Also called Egyptian millet or Egyptian grass, this species is said to have originated in southern Eurasia and was introduced into the southeastern United States as a forage crop in the 1800s. It was named after Colonel William Johnson, of Alabama, who was the first farmer to plant this species on his river bottom farm around 1840. Since then, this grass has spread globally and is considered one of the world's ten worst weeds.

Data shows that a single plant can produce 300 feet of rhizomes in a growing season as well as produce over 28,000 seeds. The thick rhizomes break off and reestablish themselves: the primary mode of dissemination is from wind and water. In Kentucky, this widespread species is one of the most troublesome in natural areas and in agricultural crops.

Johnson grass is closely related to shattercane or sorghum [*Sorghum bicolor* (L.) Moench], a variable forage sorghum that has escaped and is rarely found along roadsides at Iroquois. Sorghum is considered the fourth most important cereal after wheat, rice, and corn.

Glossary

Achene A small, dry, 1-loculed, 1-seeded fruit that does not split open; the seed attaches to the fruit wall at one place.

Adventitious roots Developing in an irregular or out-of-place position.

Adventive Spread from native range elsewhere in North America; occasional, not fully established.

Aerial roots Roots originating above the ground or water.

Alternate Not opposite, but borne at regular intervals at different levels, 1 per node.

Allelopathy Harmful chemical effect by one species upon another.

Angular Sharp-cornered.

Anther The pollen-bearing part of the stamen.

Apetalous Without petals.

Apex Tip.

Appendage A part secondary to something larger.

Appressed Lying flat, often used in reference to hairs.

Aquatic plant One that grows on, in, or under water.

Aril A fleshy basal appendage growing at or covering the seed.

Ascending Rising or curving upward.

Auricle An earlobe-shaped or clawlike lobe or appendage.

Awl-shaped Narrow and sharp-pointed; tapering upward from the base to a slender point.

Awn A sharp-pointed, bristlelike organ usually at the tip of a structure.

Axil The upper angle between a leaf or branch and the main axis.

Barbed With spinelike hooks that are usually bent backward.

Basal Pertaining to the base or foundation.

Beaked Ending in a firm, prominent point.

Berry A pulpy fruit with immersed seeds and typically a thin skin, such as a tomato.

Biennial A plant that starts from seed in the spring or summer and produces a basal rosette of leaves the first year. The following spring the overwintered plant sends up a shoot, which flowers, sets fruit, and dies in that season.

Bipinnate Divided twice.

Bladder A structure on a plant that is inflated.

Blade The expanded, flat part of a leaf or floral part.

Blunt Dull, rounded.

Bract A modified leaf.

Bristles Stiff hairs or projections from a structure.

Bulb An underground leaf bud with membranous or fleshy scales.

Bulbil A bulb-like structure that produces on aboveground parts.

Bulblet A small bulb, irrespective of origin.

Bur A seed or fruit bearing spines or prickles.

Calcareous Alkaline soil rich in calcium carbonate.

Calyx The outer whorl of flower parts; usually green; it protects the petals in the bud.

Capsule A dry dehiscent fruit composed of 2 or more carpels.

Carpel A simple pistil or a section of a compound pistil.

Catkin A scaly bracted spike or spikelike inflorescence of either all male or all female flowers.

Cauline Pertaining to the stem.

Ciliate Having a marginal fringe of hairs.

Clasping A sessile leaf with the base surrounding the stem.

Cleft Deeply cut or indented.

Cleistogamous Describing a small, closed flower that is fertilized in the bud.

Colonial Forming colonies of the same individual, usually by underground parts.

Complete A flower with sepals, petals, stamens, and pistils.

Compound leaf A leaf separated into 2 or more leaflets.

Compressed Flattened laterally.

Cone A tight cluster of scales that enclose the seeds.

Conical Cone-shaped.

Cool season grass One that actively grows in spring and fall.

Cordate Heart-shaped.

Corm A solid, thickened underground stem.

Corolla The inner whorl of the perianth; the collective name for petals.

Creeping Growing along the surface or below ground and producing roots at intervals.

Crest An elevated, irregular ridge.

Crisped Wavy or curled edges, usually referring to leaf margins.

Culm A stem of a grass or sedge.

Cultivar A variety or race that has originated and persisted under cultivation.

Deciduous Falling away, as leaves at the end of the growing season; not persistent.

Decompound More than once divided.

Decumbent Reclining on the ground but with the tip ascending.

Dehiscent Of a fruit, opening at maturity to release the seeds.

Dentate With coarse teeth or indentations that are perpendicular to the margin.

Dicot Plants usually with net-veined leaves; the perianth parts usually number 4 to 5, and the vascular bundles are in a cylindrical arrangement in the stem.

Dioecious With male and female flowers borne on separate plants.

Disjunct Separated by long distances.

Disk floret The tubular flowers in the Composite Family (*Asteraceae*).

Dissected Divided into many segments.

Divided Separated, or spreading widely.

Downy Covered with soft hairs.

Drupe A fleshy, 1-seeded fruit with a stony center.

Elliptic Oval in outline with the widest part at or about the middle.

Elongate Lengthened; stretched out.

Emergent Growing above the surface of water or soil.

Endemic Confined to a limited geographical area.

Entire Smooth-edged, the margin without projections, teeth, or lobes.

Escape A cultivated plant growing in the wild.

Ethnobotany The study of plants and people.

Even-pinnate A compound leaf ending with a pair of leaflets.

Evergreen Bearing green leaves throughout the year.

Exfoliate To peel off in shreds, plates, or layers.

Exotic Originating from outside North America.

Exserted Extending beyond a surrounding organ.

Extinct No longer in existence.

Fascicle In small clusters or bundles.

Fertile Producing fruit and seeds.

Filiform Threadlike.

Filament Any threadlike body, especially used for the part of the stamen that supports the anther.

Floret A small flower, usually one of a dense cluster.

Flower head A dense cluster of stalkless or nearly stalkless flowers.

Foliaceous Leaflike.

Follicle A dry fruit of 1 carpel that opens down 1 side only.

Frond Leaf of a fern.

Fruit A ripened ovary.

Funnelform Having the shape of a funnel.

Fusiform Spindle-shaped.

Genus A group of related plants: the first word in a scientific name.

Gland A secreting surface or structure.

Glandular Bearing glands or secreting organs.

Glaucous Covered with a whitish or bluish covering that is easily rubbed off.

Globose Shaped like a globe.

Glossy Smooth, shiny.

Glume A chaff-like bract; used particularly for one of the pair of lower bracts of a grass spikelet.

Grain A swollen, seedlike structure.

Habitat. The type of locality and immediate surroundings in which a plant grows.

Halberd-shaped Shaped like an arrowhead but with the basal lobes pointing outward at wide angles.

Head A dense cluster of flowers or fruits on a very short axis.

Herbaceous plant A vascular plant that does not develop persistent woody tissue above ground.

Herbarium A repository of dried plant specimens that have been collected from the field and stored in metal cases that document the plants of a given area, be it a county, a state, or the world. Each specimen is accompanied by a label giving the name of the plant, place collected, by whom, and when.

Hoary Covered with whitish hairs.

Horn A stiff appendage somewhat like the horn of a cow.

Husk Outermost covering of some fruits.

Hybrid A cross between 2 taxa, usually of 2 species in the same genus.

Immersed Completely submerged in water.

Incised Irregularly and deeply cut.

Indehiscent Not splitting open at maturity.

Inflated Puffed up.

Inflorescence The flowering part of a plant, especially its arrangement.

Internode The section of a stem or other structure between 2 nodes.

Introduced A plant brought in intentionally from another area for certain purposes.

Invasive plant An aggressive plant that has been introduced into an environment in which it did not evolve and has no natural enemies to limit its population.

Involucral bracts A distinct whorl of small leaves subtending a flower or an inflorescence.

Jointed Distinct nodes or joints.

Lanceolate Much longer than broad; widening above the base and tapering to the apex.

Lateral On the sides of a structure.

Latex Milky emulsion.

Lax Loose, open.

Leaflet One of the divisions of a compound leaf.

Legume A dehiscent fruit of 1 carpel that typically opens along both upper and lower sides.

Lenticel Lens-shaped dots or pits in young bark.

Ligule A hairlike or membranous projection up from the inside of a grass sheath at its junction with the blade.

Linear Long and narrow with the sides parallel.

Lip The upper or lower division of a 2-lipped corolla.

Lobe A rounded or pointed projection, larger than a tooth, from the margin of an organ.

Lobed Bearing lobes.

Mealy Flecked with another color; covered with fine granules.

Membranous Thin, papery, soft.

Mesophyte A plant that grows under medium or average conditions of moisture and light; between those in very dry and very wet environments.

Midrib The central rib of a leaf.

Monoecious With male and female flowers born separately on same plant.

Monocot Plants usually with parallel-veined leaves, the perianth parts often in arrangements of 3 or 6, and the vascular bundles scattered in the stem.

Mottled Various shades of a color or colors.

Myco-heterotrophic A plant without chlorophyll, obtaining its food from mycorhizal fungi attached to tree roots.

Native Originating from within the Central Ohio Valley.

Naturalized An exotic plant that reproduces and survives, thus becoming a part of the flora of an area.

Nerve A simple, unbranched vein or rib.

Netted Veins joining together.

Node A place on a stem where 1 or more leaves arise; a knot-like enlargement.

Notched Indented.

Nutlet A small, hard, dry nut or nutlike fruit that does not split open when mature.

Oblanceolate Narrow with the broadest part above the middle.

Oblique Unequal, such as the base of an elm leaf.

Oblong With the sides nearly parallel most of their length.

Obovate Broadest at the top and attached at the narrow end.

Obtuse Blunt or rounded at the tip.

Ocrea A membranous cylinder formed by the fusion of two stipules at a node.

Odd-pinnate A pinnately compound leaflet with a terminal leaflet.

Off-shoots Lateral shoots produced from the base of a plant.

Opposite Occurring 2 at the same level and on the opposite sides of the stem, leaves 2 per node.

Oval Broadly elliptical.

Ovary The part of a pistil that contains the ovules, which develop into the seed.

Ovate In shape like a long section of a hen's egg, with the broader end below the middle.

Ovoid Oval in outline.

Palmate Radiating from a point or center; may refer to leaves or veins.

Panicle A much-branched cluster of flowers attached at the tips of the branches.

Parasite An organism growing upon and obtaining nutrients from another; usually lacking green chlorophyll.

Pedicel The stalk of a single flower.

Peduncle A primary flower stalk supporting either a solitary flower or a flower cluster.

Peltate Attached to the center or near the center.

Pendulous Hanging or declined.

Perennial A plant that lives 3 or more years.

Perianth The floral envelope made up of the calyx and corolla, when present.

Persistent Remaining attached after like parts usually fall off.

Petal One of the inner perianth appendages of a flower, usually the colorful part of a flower.

Petiole The stalk of a leaf blade or of a compound leaf.

Pinna One of the first divisions of a pinnately compound or decompound leaf; used especially in ferns.

Pinnate A compound leaf with the leaflets on 2 opposite sides of an elongated axis.

Pith The spongy center of a stem.

Pistil The female or seed-bearing organ of a flower consisting of the ovary, style, and stigma.

Pitted Marked with small depressions.

Plumose Hairs, like a plume of a feather.

Pod A dry fruit that splits open.

Pollen The male spores in an anther.

Prickle A sharp-pointed projection growing from the surface of an organ.

Procumbent Trailing or lying on the ground, usually not rooting at the nodes.

Prostrate Lying flat.

Pubescent Covered in hairs.

Punctate Dotted with translucent or colored dots or depressions.

Raceme A simple inflorescence with stalked flowers along the central stem.

Ranked A vertical row along an axis, such as 2-ranked, 3-ranked, referring to leaves.

Ray floret The outer strap-shaped or ligulate flower of the Composite Family (*Asteraceae*).

Reclining Leaning downward.

Recurved Curved backward.

Reflexed Bent backward.

Rhizome A more or less horizontal subterranean stem usually rooting at the nodes and becoming upcurved at the tip.

Rhombic Diamond shape.

Ribbed With prominent ribs or grooves.

Root The basal part of a plant that anchors it and absorbs nutrients from the soil.

Rosette A cluster of leaves or other organs in a circular form, usually at ground level.

Rudiment An imperfectly formed, usually minute organ.

Runner A slender above ground stem rooting at the tip.

Sagittate Arrowhead shaped.

Samara A dry, winged fruit that does not split open.

Sap Juice in a plant.

Scabrous Rough to the touch.

Scalloped Wavy.

Scape A naked flowering stalk rising from the ground without proper leaves.

Secund Borne on 1 side of the axis.

Sepal One of the outer perianth appendages of a flower, usually green.

Serrate With sharp, forward-pointing teeth.

Sessile Stalkless.

Sheath A tubular envelope, usually used for that part of a grass or sedge leaf that surrounds the stem.

Shrub A woody plant that remains low, typically less than 20 feet, and usually has multiple stems.

Simple Of only 1 part, not divided or separated.

Sinus The space between 2 lobes.

Smooth Not rough.

Solitary Single.

Sorus Referring to the sporgangial clusters of ferns (*plural*, sori).

Spathe A leaflike bract sheathing or enclosing an inflorescence.

Spadix A spike with a thick, fleshy axis usually covered with imperfect flowers (those lacking either stamens or pistils).

Spatulate Spoon-shaped.

Spike A simple inflorescence with the flowers stalkless or nearly so, along a central stem.

Spikelet A floral unit, or cluster, in a grass inflorescence made up of flowers and bracts.

Spine A sharp-pointed, deep-seated outgrowth from the stem.

Spore An asexual, usually 1-celled reproductive body.

Sprawling Lying or leaning on upon or over another object.

Spur A saclike or tubular extension of a floral organ, usually with nectar glands.

Stalk The connecting or supporting part of an organ.

Stamen A male (pollen-producing) organ of a flower.

Stem Main axis of the plant.

Sterile Infertile and unproductive.

Stigma The part of the pistil that receives the pollen.

Stipule One of the paired appendages at the base of certain leaves.

Stolon An aboveground stem that lies on the ground and roots at the nodes.

Striate Marked with streaks.

Style The usually stalk-like part of a pistil between the ovary and the stigma.

Subtending Located beneath, often enclosing or embracing.

Succulent Fleshy, juicy.

Summer annual A plant that starts from seed in the spring and dies in the same year.

Suture A seam of union or separation.

Taproot A root system with a single main descending root and smaller lateral roots.

Taxa A general term for any taxonomic category regardless of its classification level.

Tendril A slender modified foliar outgrowth or stem, often coiling.

Tepal Segments of the perianth that are not clearly differentiated but look similar.

Terete Cylindrical or round.

Terminal At the tip or apex.

Ternate Arranged in 3s.

Terrestrial Living or growing in the soil.

Thorn A stiff, sharp-pointed, woody structure.

Threatened A species that is likely to disappear within a large part of its range or in all of its range.

Toothed Bearing sawlike projections along the margin.

Trailing Lying on the ground but not rooting.

Tree A woody plant with a single main trunk and a distinct and elevated crown.

Trifoliate A compound leaf with 3 leaflets, such as clovers.

Tripinnate Pinnately compound 3 times.

Tuber A thick, underground storage stem.

Tubercle A small, tuber-like body.

Tuft A cluster of hairs.

Umbel An inflorescence in which the peduncles or pedicels of a cluster spring from the same place, like ribs of an umbrella.

Unisexual Of a single sex only, either all male or all female.

Urn-shaped Round or cylindrical and contracted at or near the mouth like an urn.

Utricle A small, bladdery, 1-seeded fruit.

Waif A species not permanently naturalized, lasting a single season.

Warm season grass Grows in the warm to hot part of summer.

Webbed Covered with something woven or entangled.

Weed A plant that grows in disturbed ground and is not wanted.

Whorled In groups of 3 or more, equally spaced at the same level around the stem.

Winged A flat, membranous structure growing from the side or end of an organ.

Winter annual A plant that starts from seed in autumn develops a rosette of basal leaves before winter and then flowers and sets seed the following spring or summer.

References

Abernathy, G., ed. 2010. *Kentucky's Natural Heritage: An Illustrated Guide to Biodiversity*. The University Press of Kentucky: Lexington.

Andropogon Associates. 1994. Master Plan for Renewing Louisville's Olmsted Parks and Parkways: A Guide to Renewal and Management. Prepared for the City of Louisville, Kentucky. Louisville Olmsted Parks Conservancy, Inc., in conjunction with the Louisville and Jefferson County Parks Department. Louisville Olmsted Parks Conservancy.

Bailey, L.H. 1951. *Manual of Cultivated Plants: Most Commonly Grown in the Continental United States and Canada*. Macmillan: New York.

Barnes, T.G., and S.W. Francis. 2004. *Wildflowers and Ferns of Kentucky*. The University Press of Kentucky: Lexington.

Barnes, T.G., D. White, and M. Evans. 2008. *Rare Wildflowers of Kentucky*. The University Press of Kentucky: Lexington.

Beal, E.O., and J.W. Thieret. 1986. *Aquatic and Wetland Plants of Kentucky*. Kentucky Nature Preserves Commission Scientific and Technical Series. Number 5.

Behrendt, S., and M. Hanf. 1979. *Grass Weeds in World Agriculture*. Ludwigshafen: BASF Aktiengesellschaft.

Ben-Erick van Wyak. 2005. *Food Plants of the World: An Illustrated Guide*. Timber Press: Portland, OR.

Brandenburg, David M., and J.W. Thieret. 2003. "*Epipactis helleborine (Orchidaceae)* in Kentucky, with Overview of Literature on Biology of the Species." *Journal of the Kentucky Academy of Science* 64(1): 55–74.

Bryson, C.T., and M.S. DeFelice, eds. 2009a. *Weeds of the Midwestern United States and Central Canada*. The University of Georgia Press: Athens.

Bryson, C.T., and M.S. DeFelice, eds. 2009b. *Weeds of the South*. The University of Georgia Press: Athens.

Campbell, J.J.N., and M.E. Medley. 2012. *Atlas of Vascular Plants in Kentucky: a First Approximation*. Available at http://bluegrasswoodland.com.

Clark, R.C., and T.J. Weckman. 2008. "Annotated Catalog and Atlas of Kentucky Woody Plants." *Castanea: Occasional Papers in Eastern Botany* no. 3 (September). Allen Press: Lawrence, KS.

Cobb, B., E. Farnsworth, and C. Lowe. 2005. *A Field Guide to Ferns and Their Related Families: Northeastern and Central North America.* The Peterson Field Guide Series, 2nd edition. Houghton Mifflin Company: New York.

Cranfill, R. 1980. *Fern and Fern Allies of Kentucky.* Kentucky Nature Preserves Commission Scientific and Technical Series, Number 1.

Dalton, P.A. 1979. *Wildflowers of the Northeast in the Audubon Fairchild Garden.* Phoenix Publishing: Canaan, NH.

Dalton, P.A., and A. Novelo R. 1983. "Aquatic and Wetland Plants of the Arnold Arboretum." *Arnoldia* 43(2) 7–44.

Dirr, M.A. 1990. *Manual of Woody Landscape Plants: Their Identification, Ornamental Characteristics, Culture, Propagation and Uses.* Stipes Publishing Co: Champaign, IL.

Fernald, M.L. 1970. *Gray's Manual of Botany*, 8th ed. Van Nostrand: New York.

Flora of North America Editorial Committee, eds. 1993+. *Flora of North America North of Mexico.* 16 vols. New York and Oxford.

Foster, S., and J.A. Duke. 2000. *A Field Guide to Medicinal Plants and Herbs of Eastern and Central North America.* Peterson Field Guides. Houghton Mifflin Company: New York.

Garman, H. 1914. "Some Kentucky Weeds and Poisonous Plants." Kentucky Agricultural Experiment Station Bulletin, No. 183. State University Press: Lexington.

Gleason, H.A. and A. Cronquist. 1991. *Manual of the Vascular Plants of Northeastern United States and Adjacent Canada.* The New York Botanical Garden: New York.

Godfrey, R.K., and J.W. Wooten. 1981. *Aquatic and Wetland Plants of Southeastern United States: Dicotyledons.* The University of Georgia Press: Athens.

Godfrey, R.K., and J.W. Wooten. 1979. *Aquatic and Wetland Plants of Southeastern United States: Monocotyledons.* The University of Georgia Press: Athens.

Haragan, P.D., 1991. *Weeds of Kentucky and Adjacent States: A Field Guide.* The University Press of Kentucky: Lexington.

Heywood, V.H., 1978. *Flowering Plants of the World.* Mayflower Books: New York.

Holm, L., J. Doll, E. Holm, and J.V. Pancho. 1997. *World Weeds: Natural Histories and Distribution.* John Wiley and Sons: New York.

Jones, R.L. 2005. *Plant Life of Kentucky: An Illustrated Guide to the Vascular Flora.* The University Press of Kentucky: Lexington.

Kaufman, S.R., and W. Kaufman. 2007. *Invasive Plants: A Guide to Identification, Impacts, and Control of Common North American Species.* Stackpole Books: Mechanicsburg, PA.

Kentucky Exotic Pest Plant Council. 2012. Invasive exotic plant list at www.se-eppc.org/Ky/list.htm.

Kentucky State Nature Preserves Commission, 2010. Natural Heritage Database. Frankfort, Kentucky USA.

Kleber, John E., Editor in Chief. 2001. *The Encyclopedia of Louisville.* The University Press of Kentucky: Lexington.

Klimas, J.E., and J.A. Cunningham. 1974. *Wildflowers of Eastern America.* Alfred A. Knopf: New York.

Krochmal, A., R.S. Walters, and R.M. Doughty. 1971. *A Guide to Medicinal Plants of Appalachia.* Agricultural Handbook No. 400, Forest Service, U.S. Department of Agriculture: Washington, D.C.

Ladd, D., and F. Oberle. 2005. *Tallgrass Prairie Wildflowers: A Field Guide to Common Wildflowers and Plants of the Prairie Midwest.* Globe Pequot Press. Guilford, CT.

Lawrence, G.H.M. 1951. *Taxonomy of Vascular Plants.* Macmillan: New York.

Louisville Friends of Olmsted Parks. 1988. *Louisville's Olmsted Legacy: An Interpretive Analysis and Documentary Inventory.* City of Louisville, the Kentucky Heritage Council through Jefferson County Government and the Louisville Friends of Olmsted Parks.

Mason, Herbert L. 1957. *A Flora of the Marshes of California.* University of California Press: Berkeley.

McKinney, Landon E., and N.H. Russell. 2002. "*Violaceae* of the Southeastern United States." *Castanea* 67: 369–379.

McMurtrie, H. 1819. *Sketches of Louisville and Its Environs: Florula Louisvillensis.* Printed by S. Penn, Jun. Main-Street.

Meijer, W., 1972. *Composite Family (Asteraceae) in Kentucky.* University of Kentucky Press: Lexington.

Miller, J.H. 2003. *Nonnative Invasive Plants of Southern Forests: A Field Guide for Identification and Control.* General Technical Report SRS-62. United States Department of Agriculture, United States Forest Service, Southern Research Station, Ashville, NC.

Moerman, D.E. 1998. *Native American Ethnobotany.* Timber Press: Portland, OR.

Moore, R.J., and C. Frankton. 1974. *The Thistles of Canada.* Research Branch, Canada Department of Agriculture, Monograph #10.

Muenscher. W.C. 1955. *Weeds.* Macmillan: New York.

Olmsted Associates, Inc. 1974. Journal of the Development of Cherokee Park. Louisville, Kentucky, 1891–1974. Brookline, MA.

Peattie, D.C. 1956. *A Natural History of Trees of Eastern and Central North America,* 2nd ed. Bonanza Books: New York.

Polunin, O. 1969. *Flowers of Europe: A Field Guide.* Oxford University Press: London.

Randall, J.M., and J. Marinelli, ed. 1996. *Invasive Plants: Weeds of the Global Garden.* Handbook #149. Science Press: Brooklyn, New York.

Slack, M. 1941. "A Survey of the Flora of Cherokee Park at Louisville, Kentucky." Unpublished master's thesis, Cornell University: Ithaca.

Spencer, E.R. 1940. *Just Weeds.* Scribner's: New York.

Sterry, P. 2006. *Complete Guide to British Wildflowers.* HarperCollins Publishers: London.

Wharton, M.E., and R.W. Barbour. 1973. *Trees and Shrubs of Kentucky.* The University Press of Kentucky: Lexington.

Wharton, M.E., and R.W. Barbour. 1971. *Wildflowers and Ferns of Kentucky.* The University Press of Kentucky: Lexington.

Westbrooks, R. 1998. *Invasive Plants, Changing the Landscape of America: Fact Book.* Federal Interagency Committee for the Management of Noxious and Exotic Weeds, Washington, D.C.

Williams, M.D. 2007. *Identifying Trees: An All-Season Guide to Eastern North America.* Stackpole Books: Mechanicsburg, PA.

Woods, Michael. 2008. "The Genera *Desmodium* and *Hylodesmum* (Fabaceae) in Alabama." Castanea 73(1): 46–69.

Online Sources

Biota of North America Program:
www.bonap.org

Flora of North America:
www.efloras.org

Frederick Law Olmsted Parks:
www.olmstedparks.org

Illinois Wildflowers:
www.illinoiswildflowers.info

Illustrated Atlas of Vascular Plants in Kentucky: A First Approximation:
http://bluegrasswoodland.com

Kentucky Division of Forestry:
www.forestry.ky.gov

Kentucky Exotic Pest Plant Council:
www.se-eppc.org/ky/

Kentucky Native Plant Society:
www.knps.org

Kentucky Rare Plant Database:
www.eppcapp.ky.gov/nprareplants/index.aspx

King County Noxious Weed List:
www.kingcounty.gov/environment/animalsAndplants/noxious-weeds/laws/list.aspx

Metro Parks, City of Louisville:
www.louisvilleky.gov/metroparks

Missouri Flora:
www.missouriplants.com

Plant Conservation Alliance's Alien Plant Fact Sheets:
www.nps.gov/plants/alien/factmain.htm

Acknowledgments

There are several people who have been instrumental while working on this book. First and foremost, I am indebted to my husband, Chris, who has been supportive of this project every step of the way. I am also indebted to Alan Nations, arborist, naturalist, founder of NativeScapes, Inc., and former Naturalist and Restoration Specialist with the Louisville Olmsted Parks Conservancy. His passion, knowledge, and dedication to restoring the health of these great urban parks prompted the start-up of the 2005 Woodlands Restoration campaign. I am very grateful to you for recognizing the need, and value, of cataloging the plants: a mere snapshot in time!

To David Fothergill, former Landscape Supervisor for Metro Parks and field buddy since day one. I could not have asked for a more enthusiastic, knowledgeable, and kind person to botanize with.

To Susan Rademacher, Director of the Pittsburg Parks Conservancy and internationally known Olmsted scholar, it has been an honor to have worked with you as former Executive Director of the Louisville Olmsted Parks Conservancy.

To the photographers, Susan Wilson and Chris Bidwell, both trauma nurses by profession and naturalists by passion, this book is a diary of our days in the field exploring, well captured by your exceptional photographs.

Special thanks go to Major Waltman, Research Specialist for the Louisville Olmsted Parks Conservancy; Mimi Zinniel, President/CEO of the Louisville Olmsted Parks Conservancy, and to Dan Jones, Chairman and CEO, 21st Century Parks, for your support and valuable input.

As with any floristic project, time becomes a limiting factor, and I could not have accurately written this book without relying on the vast knowledge of several key people who work with the Louisville Olmsted Parks Conservancy or

Metro Parks currently or have done so in the past. Thanks to Josh Wysor, John Swintosky, Liz Mortenson, Sarah Wolff, Katie Greene, Adrian Camacho, Carl Suk, Robert Woodford, Matt Spalding, and Andrew Oost.

To these professional mentors: Ron Jones, Rob Paratley, Julian Campbell, Margaret Carreiro, Larry Alice, Ross Clark, Deborah White, Joyce Bender, Allen Bush, and Landon McKinney, I am sincerely thankful for your expertise and advice in editing my manuscript, looking at plants I questioned, or spending time in the field or herbarium with me. This scholarly input, your friendship, and encouragement has inspired and guided me through this venture. Any mistakes it still may have are my responsibility.

And, lastly, to Mabel Slack, John Thieret, Harrison Garman, and other botanists, both past and present: here's to you, a rare breed whose passion for exploring the field, collecting, and documenting plants in Kentucky has laid down the ground work so that others may build upon and better understand the "pristine" flora of long ago as well as the changing landscape of today. Preserving what is left of our rich botanical heritage, both in the field and herbaria, depends on this timeless knowledge. I am forever grateful!

Special Thanks

I would like to thank the field crews with the Louisville Olmsted Parks Conservancy and Metro Parks with whom I have worked with since 2005. The endless hours spent physically working in the relentless heat or bitter cold has not gone unnoticed. You are the backbone of these parks.

Photographers' Note

As the photographers for this book, we would first and foremost like to thank Pat Haragan for her trust and patience with two amateurs. We have benefited from her vast knowledge of the flora that surrounds us in our wonderful park system in Louisville. As trauma nurses by trade and naturalists by choice, our great love of nature, especially our wildflowers, has truly been expanded by our days in the field photographing and identifying the flora around us. Thanks to all our families for their love and support during this project. We also want to thank our dear friend Dr. Tom Barnes from the University of Kentucky, who without his expertise and positive feedback these pictures would not have been possible. All photographs were taken with our Nikon D300 camera using either an AF micro Nikkor 200mm or an AFS Nikkor 24–120mm lens.—Chris Bidwell and Susan Wilson

Olmsted Parks Conservancy

The mission of Louisville's Olmsted Parks Conservancy is to enrich the life of everyone in the community by restoring, enhancing, and preserving the Frederick Law Olmsted Parks and Parkways—Louisville's great natural and recreational assets.

In one way or another, parks have the ability to improve almost every aspect of life, just as Frederick Law Olmsted had envisioned 120 years ago. Parks enhance the environmental, economic, recreational, and social health of the entire community.

Louisville's Olmsted Parks need active and ongoing restoration to enhance their habitat value for native plants and animals, and to preserve air and water quality. Caring for these historic treasures and seeing that they remain valuable assets for the community is the heart of the work undertaken by Olmsted Parks Conservancy.

Volunteering with Olmsted Parks Conservancy is a hands-on learning experience with the satisfaction of immediate, tangible results. Volunteers eradicate invasive plants from the park woodlands; they plant native species of trees and shrubs; they keep the forests healthy and the trails sustainable. Volunteering is an excellent way to connect with nature and the community while improving the livability of Louisville now, and for future generations.

It takes hard work, funding, and community interest to preserve park resources. Louisville Olmsted Parks Conservancy leads the movement to enhance and restore these resources, and invites the community to realize its stake in the future of our Frederick Law Olmsted Parks by volunteering to help with this important work.

Olmsted Parks Conservancy, 1299 Trevilian Way, Louisville KY 40213
Phone: 502-456-8125
info@olmstedparks.org
www.olmstedparks.org

Index

Acalypha
  rhomboidea, **310**
  virginica, 310
Acer
  negundo, **384**
  rubrum, 379
  saccharium, **378**
  saccharum, **379–80**
Actaea
  alba, **122**
  pachypoda, **122**
Aesculus glabra, **387**
Agalinis
  besseyana, 230
  purpurea, **230**
  tenuifolia, 230
Ageratina altissima, **90**
Agrimonia pubescens, **193**
Ailanthus altissima, **376–77**
Akebia. *See* Chocolate vine
Akebia quinata, **275**
Alehoof. *See* Ground-ivy
Alliaria petiolata, **49–50**
Allium
  burdickii, **82**
  tricoccum, 82
  vineale, **208**
Amaranthus
  hybridus, **294–95**
  retroflexus, 294,
    295

Ambrosia
  artemisiifolia, 300
  trifida, **300–301**
Amelanchier arborea, **360–61**
American beech, **343**
American bittersweet, 304,
  305
American elm, 369
American false pennyroyal,
  272
American groundnut, **268**
American heal-all. *See*
  Self-heal
American hogpeanut, **269**
American lime. *See* Basswood
American lotus, **113**
American plantain, 329
American water-willow, **254**
Ampelamus albidus, **89**
Ampelopsis brevipedunculata,
  321–22
Amphicarpaea bracteata, **269**
Amur honeysuckle, **97–98**
Andropogon
  gerardii, **392–93**
  virginicus, 392
Anemonella thalictroides,
  72, 73
Anemone virginiana, **123**
Angel-eyes. *See* Spring bluets
Annual fleabane, 45

Antennaria plantaginifolia,
  44
Apios americana, **268**
Apocynum cannabinum, **88**
Appendaged waterleaf. *See*
  Biennial waterleaf
Aquilegia canadensis, **206**
Arabis laevigata, **51**
Arisaema
  dracontium, **287**
  triphyllum, **288**
Artemisia vulgaris, **302–3**
Arundinaria gigantea, **394**
Asarum canadense, **328**
Asclepias incarnata, **209**
Ascyrum hypericoides, **184**
Ashy beardstongue, 132
Ashy sunflower, **163**
Asian bittersweet, **304**
Asiatic bittersweet. *See* Asian
  bittersweet
Asimina triloba, **337–38**
Asplenium
  platyneuron, **33**
  rhizophyllum, **34**
Aster
  cordifolius, **283**
  laevis, **282**
  novae-angliae, **284**
  pilosus, **138**
  racemosus, 138

Aster (cont.)
   solidagineus, **94**
   vimineus, 138
Aster
   common blue heart-leaved, **283**
   narrow-leaved white-top, **94**
   New England, **284**
   old field, **138**
   small-headed, 138
   smooth, **282**
Asthma weed. *See* Indian tobacco
Avens
   spring, **158**
   white, **126**
Axillary goldenrod, **197**

Bamboo. *See* Giant cane
Bamboo honey. *See* Japanese knotweed
Baptisia
   alba, **106**
   lactea, **106**
   leucantha, **106**
Basswood, **366**
Beaked corn-salad, 78
Beaked hawkweed, **164**
Beardstongue
   ashy, 132
   foxglove, **132–33**
Beech
   American, **343**
   blue. *See* Ironwood
   water. *See* Ironwood
Beechdrops, **334**
Beefstake plant. *See* Perilla mint
Bellwort
   largeflower, **160**
   perfoliate, 160
Bent trillium, 77
Bermuda grass, **396**
Bidens
   bipinnata, 161
   frondosa, **161**
Biennial waterleaf, **238**

Big bluestem, **392–93**
Bindweed
   field, 102
   hedge, **102**
Birdseye speedwell, **250–51**
Bitternut hickory, **372**
Bittersweet
   American, 304, **305**
   Asian, **304**
   round-leaved. *See* Asian bittersweet
   weedy, 305
Black cohosh, **124**
Black-eyed Susan, 170
   cutleaf, **169**
Blackjack oak, **347**
Black locust, **371**
Black medic, **181**
Black nightshade, 279
Black oak, 350
Black raspberry, 75
Black walnut, **375**
Bladdernut, **386**
Blazing-star
   plains, 214
   sessile, **214–15**
Bloodroot, **71**
Bloody butchers. *See* Toadshade trillium
Blue beech. *See* Ironwood
Blue-eyed lily. *See* Narrowleaf blue-eyed grass
Bluet(s)
   broad-leaved, **277**
   small, 248
   spring, **248–49**
   summer. *See* Broad-leaved bluets
Bluevine. *See* Sandvine
Border privet, 114
Botrychium virginianum, **39**
Bottlebrush grass, **399**
Boxelder maple, **384**
Brassica rapa, **143**
Breast-weed. *See* Lizard's tail
Bristly greenbrier, 318
Broadleaf dock, 290–91, 316
Broad-leaved bluets, **277**

Broad-leaved toothwort, 54
Broom-edge, 392
Broom hickory. *See* Pignut hickory
Brown-eyed Susan. *See* Three-lobed coneflower
Buckbush. *See* Coralberry
Buckthorn
   Dahurian, 382
   European, **382–83**
Bull thistle, **210–11**
Bundleflower, **107**
Bur cucumber, **309**
Burdick's wild leek, **82**
Bur oak, **346**
Bursting heart. *See* Hearts-a-bursting-with-love
Bushy seedbox. *See* Square-pod water-primrose
Buttercup
   fig. *See* Lesser celandine
   hairy, **192**
   hooked, **155**
Butter-weed. *See* Horseweed
Buttonball-tree. *See* Sycamore
Button snake-root. *See* Sessile blazing-star
Buttonweed
   rough, 129
   Virginia, **129**
Buttonwood. *See* Sycamore

Cabinet cherry. *See* Wild black cherry
Calystegia sepium, **102**
Camassia scilloides, **236**
Campanula americana, **264**
Campanulastrum americanum, **264**
Campis radicans, **178**
Canada germander, **223–24**
Canada thistle, **258–59**
Cancer-drop. *See* Beechdrops
Cancer-root. *See* Beechdrops
Cancer root. *See* Wild sage
Cancerweed. *See* Wild sage
Candlewicks. *See* Common mullein

Cane reed. *See* Giant cane
Capsella bursa-pastoris, **52**
Cardamine hirsuta, **53**
Cardinal-flower, **216–17**
Carduus nutans, 210, 211
Carex, **388**
Carex albursina, **388**
Carolina cranes-bill, 202
Carolina elephant's-foot, **261**
Carolina poplar. *See* Eastern
    cottonwood
Carolina wild petunia, 255
Carpinus caroliniana, **339**
Carya
    cordiformis, **372**
    glabra, **373**
    ovata, **374**
Cassia fasciculata, **180**
Cats paw. *See* Hearts-
    a-bursting-with-love
Caulophyllum thalictroides,
    289
Celandine poppy, **151**
Celastrus
    orbiculatus, **304**
    scandens, 304, **305**
Celtis occidentalis, **367**
Cerastium vulgatum, **60**
Cercis
    canadensis, **342**
    siliquastrum, 342
Chaerophyllum procumbens,
    42
Chamaecrista fasciculata, **180**
Chasmanthium latifolium,
    395
Chenopodium album, **308**
Chestnut oak, **348–49**
Chicken-weed, 61
Chickweed
    common, **61**
    common mouse-ear, **60**
    great, **62**
    Indian-, 61
Chicory, **256–57**
Chinese privet, **114–15**
Chinese yam, 104
Chinquapin oak, 348

Chocolate vine, 19, **275**
Christmas fern, **37**
Chufa, 389
Cichorium intybus, **256–57**
Cimicifuga racemosa, **124**
Cirsium
    arvense, **258–59**
    vulgare, **210–11**
Citronella. *See* Northern
    horse-balm
Claytonia virgnica, **205**
Clearweed, **319**
Cleavers, **76**
Clematis
    terniflora, **125**
    virginiana, 125
Climbing euonymus, **307**
Climbing nightshade, **281**
Climbing spindle berry. *See*
    Asian bittersweet
Clover
    red, **109**
    running buffalo, 18
    white, **109**
Cluster sanicle, **299**
Cock's foot. *See* Orchard
    grass
Coffeebean. *See* Kentucky
    coffeetree
Coffeenut. *See* Kentucky
    coffeetree
Collinsonia canadensis, **186**
Colonial dwarf-dandelion,
    165
Commelina
    communis, **266**
    diffusa, 266
Common arrowhead, **81**
Common barnyard grass,
    398
Common blackberry, **75**
Common blue cohosh, **289**
Common blue heart-leaved
    aster, **283**
Common blue violet, **253**
Common cat-tail, **332–33**
Common chickweed, **61**
Common dandelion, **142**

Common dayflower, **266**
Common elderberry, **99**
Common evening-primrose,
    190
Common false foxglove, 230
Common goldenrod, **198–99**
Common grape fern, **39**
Common greenbrier, **318**
Common hackberry, **367**
Common marsh-pink, **221**
Common mouse-ear
    chickweed, **60**
Common mullein, **194–95**
Common periwinkle, **232**
Common privet, 114, 115
Common quickweed, **93**
Common ragweed, 300
Common scouring-rush, **38**
Common speedwell, 252
Common St. John's-wort,
    179
Common yellow wood-sorrel,
    149–50
Coneflower
    cutleaf. *See* Cutleaf
        black-eyed Susan
    gray-headed. *See* Yellow
        coneflower
    green-headed. *See* Cutleaf
        black-eyed Susan
    purple, **212**
    thin-leaved. *See* Three-
        lobed coneflower
    three-lobed, **170–71**
    yellow, **168**
Conium maculatum, **83**
Conoclinium coelestinum, **260**
Convolvulus arvensis, 102
Conyza canadensis, **91**
Copperleaf
    rhombic, **310**
    Virginia, 310
Coralberry, **100**
Cork elm. *See* Winged elm
Corn-salad
    beaked, 78
    navel, 78
Cornus florida, **381**

Coronilla varia, **218**
Corydalis flavula, **145**
Cow-itch. *See* Trumpet creeper
Cowtail. *See* Horseweed
Coyote willow. *See* Sandbar willow
Crabweed, **312**
Crane-fly orchid, **331**
Cream violet, **80**
Creeping Charlie. *See* Ground-ivy; Moneywort
Creeping dayflower, 266
Creeping Jenny. *See* Moneywort
Creeping water-primrose, **189**
Creeping yellow cress, **144**
Crested coral-root, **330**
Crowfoot
  hairy small-flowered, **154**
  small-flowered, 154
Crown-vetch, **218**
Cryptotaenia canadensis, **84**
Cunila origanoides, **286**
Cup-plant, **172**
Curly dock, **290–91**, 316
Cutleaf black-eyed Susan, **169**
Cutleaf coneflower. *See* Cutleaf black-eyed Susan
Cynanchum laeve, **89**
Cynodon dactylon, **396**
Cynoglossum virginianum, **233**
Cyperus
  esculentus, 389
  strigosus, **389**
Cystopteris protrusa, **35**

Dactylis glomerata, **397**
Dahurian buckthorn, 382
Dandelion
  colonial dwarf-, 165
  common, **142**
  orange dwarf-, **165**

Daucus carota, **85–86**
Dayflower
  common, **266**
  creeping, 266
Decumbent five-fingers. *See* Old field five-fingers
Deerberry, 105
Deer potato. *See* Sessile blazing-star
Deerwood. *See* Hop-hornbeam
Delphinium tricorne, **246**
Dentaria
  diphylla, 54
  laciniata, **54**
Desmanthus illinoensis, **107**
Desmodium
  canescens, **270**
  glabellum, **219**
  nudiflorum, 219
  paniculatum, **219**
  pauciflorum, 270
  perplexum, **219**
Devil's beggar-ticks, **161**
Devil's-clothesline. *See* Common greenbrier
Devil's shoelaces. *See* Trumpet creeper
Dianthera americana, **254**
Dicentra
  canadensis, **66**
  cucullaria, **67**
Diodia
  teres, 129
  virginiana, **129**
Dioscorea
  polystachya, 104
  villosa, **104**
Diospyros virginiana, **341**
Divaricate sunflower, **162**
Dock
  broadleaf, 290–91, 316
  curly, **290–91**, 316
  pale, **316**
  smooth. *See* Pale dock
  water. *See* Pale dock
Dock-leaved smartweed, **119**

Dog-tongue. *See* Wild comfrey
Dogtooth violet. *See* Yellow fawn-lily
Doll's eye. *See* White baneberry
Dotted smartweed, **120**
Downy serviceberry, **360–61**
Downy trailing lespedeza, 220
Draba
  brachycarpa, 55, 56
  verna, **55–56**
Dragon root. *See* Green-dragon
Dragon-tail. *See* Green-dragon
Dryland blueberry. *See* Low-bush blueberry
Duchesnea indica, **156**
Duck potato. *See* Common arrowhead
Dutchman's-breeches, **67**
Dwarf crested iris, **240**
Dwarf-dandelion
  colonial, 165
  orange, **165**
Dwarf larkspur, **246**

Early meadow-rue, **292**
Earth almond. *See* Chufa
Eastern cottonwood, **363–64**
Eastern figwort, **231**
Eastern red-cedar, **335**
Ebony spleenwort, **33**
Echinacea purpurea, **212**
Echinochloa crus-galli, **398**
Egyptian grass. *See* Johnson grass
Egyptian millet. *See* Johnson grass
Eichhornia crassipes, 276
Elephantopus carolinianus, **261**
Elm
  American, 369
  cork. *See* Winged elm
  red, **369–70**
  winged, **368**

Elm-leaved goldenrod, **200**
Elymus hystrix, **399**
Enemion biternatum, **73**
English ivy, **326–27**
English plantain, **329**
Epifagus virginiana, **334**
Epipactis helleborine,
　　313–14
Equisetum hyemale, **38**
Erechtites hieracifolia, **92**
Erigenia bulbosa, **43**
Erigeron
　　annuus, 45
　　philadelphicus, **45–46**
Eryngium yuccifolium, **87**
Erythronium
　　albidum, **70**
　　americanum, **148**
Eulalia viminea, **400–401**
Euonymus
　　americanus, **306**
　　fortunei, **307**
Eupatorium
　　coelestinum, **260**
　　fistulosum, 213
　　purpuream, **213**
　　rugosoum, **90**
　　serotinum, **136**
European buckthorn,
　　382–83

Fagus grandifolia, **343**
False foxglove
　　common, 230
　　smooth, **230**
False rue-anemone, **73**
False Solomon's-seal, **63**
Fatoua villosa, **312**
Fawn-lily
　　white, **70**
　　yellow, **148**
Fern
　　Christmas, **37**
　　common grape, **39**
　　sensitive, **36**
　　Southern beech, **41**
　　southern bladder, **35**
　　walking, **34**

Fetid buckeye. *See* Ohio
　　buckeye
Fetidshrub. *See* Pawpaw
Few flowered tick-trefoil,
　　270
Field bindweed, 102
Field garlic, **208**
Field hedge-parsley, 42
Field mustard. *See* Turnip
Field pansy, **79**
Field rush. *See* Path rush
Fig buttercup. *See* Lesser
　　celandine
Fireweed, **92**
Five-fingers
　　old field, 157
　　running, **157**
Five-leave akebia. *See*
　　Chocolate vine
Five-parted toothwort, **54**
Flannel-leaf. *See* Common
　　mullein
Flathead oats. *See* River wood
　　oats
Fleabane
　　annual, 45
　　Philadelphia, **45–46**
Fleeceflower. *See* Japanese
　　knotweed
Flowering dogwood,
　　381
Forest phlox, **244**
Foxglove
　　common false, 230
　　smooth false, **230**
Foxglove beardstongue,
　　132–33
Foxtail
　　giant, 403
　　yellow, **403**
Fragaria virginiana, **156**
Fraxinus
　　americana, **385**
　　biltomoreana, **385**
Fringed quickweed. *See*
　　Common quickweed
Frogfruit, **134**
Frostweed, 201

Galinsoga
　　ciliata, **93**
　　quadriradiata, **93**
Galium
　　aparine, **76**
　　circaezens, **317**
　　concinnum, **130**
Garlic-mustard, **49–50**
Garlic penny-cress, **57–58**
Geranium
　　carolinianum, 202
　　maculatum, **202**
Geum
　　canadense, **126**
　　vernum, **158**
Ghost corn. *See* Squirrel-corn
Giant cane, **394**
Giant foxtail, 403
Giant ragweed, **300–301**
Gillenia stipulata, **127**
Gill-over-the-ground. *See*
　　Ground-ivy
Ginger
　　green. *See* Mugwort
　　wild, **328**
Glechoma hederacea, **241**
Golden-carpet, 65
Goldenglow. *See* Cutleaf
　　black-eyed Susan
Goldenrod
　　axillary, **197**
　　common, **198–99**
　　elm-leaved, **200**
　　hairy, **174**
　　smooth, 198
　　wreath. *See* Axillary
　　　goldenrod
　　zigzag, **173**
Goldenseal, **74**
Goosegrass. *See* Cleavers
Grass
　　Bermuda, **396**
　　bottlebrush, **399**
　　common barnyard, **398**
　　Egyptian. *See* Johnson
　　　grass
　　Indian, **404–5**
　　Johnson, **406–7**

Grass (cont.)
  narrowleaf blue-eyed,
    271
  orchard, **397**
  packing. *See* Nepalese
    eulalia
  star-eyed. *See* Narrowleaf
    blue-eyed grass
  switch, **402**
Gravelroot. *See* Northern
    horse-balm
Gray-headed coneflower. *See*
    Yellow coneflower
Great blue lobelia, **285**
Great chickweed, **62**
Greenbrier
  bristly, 318
  common, **318**
Green-dragon, **287**
Green ginger. *See* Mugwort
Green-headed coneflower. *See*
    Cutleaf black-eyed
    Susan
Green violet, **293**
Ground-ivy, **241**
Gymnocladus dioicus, **370**

Hairy bitter-cress. *See* Hoary
    bitter-cress
Hairy buttercup, **192**
Hairy goldenrod, **174**
Hairy lespedeza, **108**
Hairy small-flowered
    crowfoot, **154**
Hairy waterleaf, 238
Harbinger-of-spring, **43**
Hearts-a-bursting-with-love,
    306
Hedeoma pulegioides, 272
Hedera helix, **326–27**
Hedge-apple. *See* Osage-
    orange
Hedge bindweed, **102**
Helianthus
  divaricatus, **162**
  mollis, **163**
  tuberosus, **196**
Helleborine, **313–14**

Hell-ropes. *See* Common
    greenbrier
Hell vine. *See* Trumpet
    creeper
Hemp dogbane, **88**
Henbit, 203–4
Hepatica acutiloba, **247**
Hexalectris spicata, **330**
Hibiscus
  laevis, 111
  militaris, 111
  mosc…tos, **111–12**
Hickory
  bitternut, 372
  broom. *See* Pignut hickory
  pignut, 373
  scaly-barked. *See* Shagbark
    hickory
  shagbark, 374
  shellbark. *See* Shagbark
    hickory
  smoothbark. *See* Pignut
    hickory
  swamp. *See* Pignut hickory
  upland. *See* Shagbark
    hickory
Hieracium gronovii,
    164
Hoary bitter-cress, **53**
Hoary tick-trefoil, **270**
Hogweed. *See* Horseweed
Hollow-stemmed joe-pye
    weed, 213
Honesty, **235**
Honewort, **84**
Honeysuckle
  amur, **97–98**
  Japanese, **59**
Honeyvine milkweed. *See*
    Sandvine
Honeyvine swallowort. *See*
    Sandvine
Hooked buttercup, **155**
Hop-hornbeam, **340**
Hopniss. *See* American
    groundnut
Horse-nettle, **279–80**
Horseweed, **91**

Hound's-tongue. *See* Wild
    comfrey
Houstonia
  caerulea, **248–49**
  lanceolata, **277**
  purpurea, **277**
  pusilla, 248
Hyacinth
  water, 276
  wild, **236**
Hybanthus concolor, **293**
Hydrangea arborescens, **110**
Hydrastis canadensis, **74**
Hydrophyllum
  appendiculatum, **238**
  canadense, **69**
  macrophyllum, 238
Hypericum
  perforatum, **179**
  punctatum, 179
  stragulum, **184**
Hypoxis hirsuta, **146**
Hystrix patula, **399**

Illinois wood-sorrel, 149
Impatiens
  capensis, **176–77**
  pallida, 176
Indian-chickweed, 61
Indian grass, **404–5**
Indian gravelroot. *See*
    Purple-node joe-pye
    weed
Indian pipe, **139**
Indian potato. *See* American
    groundnut
Indian strawberry, **156**
Indian tobacco, **265**
Indian turnip. *See* Jack-in-
    the-pulpit
Innocence. *See* Spring bluets
Iodanthus pinnatifidus, **263**
Ipomoea
  batatus, 103
  pandurata, **103**
Iris
  cristata Soland, **240**
  pseudacorus, **185**

Iris
    dwarf crested, **240**
    yellow flag, **185**
Ironwood, **339**. *See also*
    Hop-hornbeam
Isopyrum biternatum, **73**

Jack-in-the-pulpit, **288**
Jacob's ladder, **245**
Jacob's staff. *See* Common
    mullein
Japanese chaff flower, 19
Japanese honeysuckle, **59**
Japanese knotweed, **315**
Japanese privet, 114
Jeffersonia diphylla, **46**
Jersualem-artichoke, **196**
Joe-pye weed
    hollow-stemmed, 213
    purple-node, 213
Johnson grass, **406–7**
Judas tree, 342
Juglans nigra, **375**
Jumpseed, **121**
Juncus tenuis, **390**
Juneberry. *See* Downy
    serviceberry
Juniperus virginiana,
    335
Justicia americana, **254**

Kentucky coffeetree, **370**
Kidneyroot. *See* Purple-node
    joe-pye weed
Krigia
    biflora, **165**
    dandelion, 165

Lactuca
    floridana, **262**
    serriola, **166**
Lamb's-quarters, **308**
Lamium
    amplexicaule, 203–4
    galeobdolon, **187**
    purpureum, 203–4
Laportea canadensis, **320**
Largeflower bellwort, **160**

Large-flowered sensitive-pea.
    See Partridge-pea
Large houstonia. *See*
    Broad-leaved bluets
Largeleaf wild indigo, **106**
Late eupatorium, **136**
Leek
    Burdick's wild, **82**
    narrowleaf wild. *See*
    Burdick's wild leek
    wild, 82
Lespedeza
    hirta, **108**
    procumbens, 220
    repens, **220**
Lespedeza
    downy trailing, 220
    hairy, **108**
    smooth trailing, **220**
Lesser celandine, **152–53**
Lettuce
    prickly, **166**
    woodland, **262**
Leverwood.
    See Hop-hornbeam
Liatris
    spicata, **214–15**
    sqarrosa, 214
Ligustrum
    obtusifolium, 114
    sinense, **114–15**
    vulgare, 114, 115
Limestone wild petunia, **255**
Limetree. *See* Basswood
Linden. *See* Basswood
Lindera benzoin, **147**
Liquidambar styraciflua, **352**
Liriodendron tulipifera,
    354–55
Lizard's tail, **131**
Lobelia
    cardinalis, **216–17**
    inflata, **265**
    siphilitica, **285**
Lonicera
    japonica, **59**
    maackii, **97–98**
Lopseed, **227**

Low-bush blueberry, **105**
Ludwigia
    alternifolia, **188**
    peploides, **189**
Lunaria annua, **235**
Luzula echinata, **391**
Lyreleaf sage. *See* Wild sage
Lysimachia nummularia, **191**
Lythrum salicaria, **225–26**

Maclura pomifera, **356**
Maianthemum racemosum, **63**
Maple
    boxelder, **384**
    red, 379
    silver, **378**
    sugar, **379–80**
Mapleleaf viburnum, **101**
Maple-leaved waterleaf, **69**
Mares-tail. *See* Horseweed
Matricaria
    discoidea, **167**
    matricarioides, **167**
May-apple, **47–48**
Medicago lupulina, **181**
Menispermum candense, **311**
Mertensia virginica, **234**
Miami-mist, **239**
Microstegium vimineum,
    400–401
Midwestern Indian-physic,
    127
Mint
    perilla, **273**
    stone, **286**
Mist-flower, **260**
Monarda fistulosa, **222**
Money plant. *See* Honesty
Moneywort, **191**
Monotropa uniflora, **139**
Moonseed, **311**
Moonwort. *See* Honesty
Morus
    alba, **357–58**
    rubra, 357, 358
Moth mullein, 194
Mugweed. *See* Mugwort
Mugwort, **302–3**

Mulberry
  red, 357, 358
  white, **357–58**
Mullein
  common, **194–95**
  moth, 194
Multiflora rose, **128**
Musk thistle, 210, 211

Naked tick-trefoil, 219
Nap-at-noon. *See* Star-of-
  Bethlehem
Narrowleaf blue-eyed grass,
  271
Narrowleaf wild leek. *See*
  Burdick's wild leek
Narrowleaf willow. *See*
  Sandbar willow
Narrow-leaved white-top
  aster, **94**
Nasturtium officinale, **95–96**
Navel corn-salad, **78**
Nelumbo lutea, **113**
Nepalese eulalia, **400–401**
New England aster, **284**
Nightshade
  black, 279
  climbing, **281**
Nimbleweed. *See* Tall
  anemone
Nonesuch. *See* Black medic
Northern horse-balm, **186**
Northern nutsedge. *See*
  Chufa
Northern red oak, **350–51**
Northern wild senna, **182**

Oak
  black, 350
  blackjack, **347**
  bur, **346**
  chestnut, **348–49**
  Chinquapin, 348
  Northern red, **350–51**
  post, 344, 345
  rock. *See* Chestnut oak
  rock chestnut. *See*
    Chestnut oak

  Shumard, 350–51
  white, **344–45**
Oenothera biennis, **190**
Ohio buckeye, **387**
Old field aster, **138**
Old field five-fingers, 157
Onoclea sensibilis, **36**
Ophioglossum vulgatum, **40**
Orange dwarf-dandelion, **165**
Orange-root. *See* Goldenseal
Orange touch-me-not,
  176–77
Orchard grass, **397**
Orchid
  crane-fly, **331**
  ragged fringed, **116**
Oriental staff vine. *See* Asian
  bittersweet
Ornithogalum umbellatum,
  68
Osage-orange, **356**
Ostrya virginiana, **340**
Oxalis
  Illinoensis, 149
  stricta, **149–50**
  violacea, **243**

Packera
  glabella, **140**
  obovata, **141**
Packing grass. *See* Nepalese
  eulalia
Pale corydalis, **145**
Pale dock, **316**
Pale-flowered leaf-cup, **137**
Panicled tick-trefoil, **219**
Panicum virgatum, **402**
Parthenocissus quinquefolia,
  325
Partridge-pea, **180**
Path rush, **390**
Pawpaw, **337–38**
Pencil-flower, **183**
Pennsylvania smartweed,
  228–29
Penstemon
  canescens, 132
  digitalis, **132–33**

Pepper turnip. *See* Jack-in-
  the-pulpit
Perfoliate bellwort, 160
Perfoliate penny-cress, 57
Perilla frutescens, **273**
Perilla mint, **273**
Persimmon, **341**
Peruvian daisy. *See* Common
  quickweed
Petunia
  Carolina wild, 255
  limestone wild, **255**
Phacelia purshii, **239**
Phegopteris hexagonoptera, **41**
Philadelphia fleabane,
  45–46
Phlox divaricata, **244**
Phryma leptostachya,
  227
Phyla lanceolata, **134**
Phytolacca americana, **118**
Pickerel-weed, **276**
Pigeon grape. *See* Summer
  grape
Pignut hickory, **373**
Pigweed
  redroot. *See* Rough
    pigweed
  rough, 294, 295
  smooth, **294–95**
Pilea pumila, **319**
Pineapple-weed, **167**
Pinus strobus, **336**
Plains blazing-star, **214**
Plane tree. *See* Sycamore
Plantago
  lanceolata, **329**
  rugelii, 329
Plantain
  American, 329
  English, **329**
Plantain pussytoes, **44**
Platanthera lacera, **116**
Platanus occidentalis, **359**
Podophyllum peltatum, **47–48**
Poison hemlock, **83**
Poison-ivy, **298**
Pokeweed, **118**

Polemonium reptans, **245**
Polygonatum
  biflorum, 63, **64**
  commutatum, **66**
Polygonum
  cuspidatum, **315**
  lapathifolium, **119**
  pensylvanicum, **228–29**
  persicaria, 228, 229
  punctatum, **120**
  virginianum, **121**
Polymnia canadensis, **137**
Polystichum acrostichoides, **37**
Pontederia cordata, **276**
Poorjoe, 129
Populus deltoides, **363–64**
Porcelain-berry, **321–22**
Porteranthus stipulatus, **127**
Post oak, 344, 345
Potato bean. *See* American
  groundnut
Potato dandelion. *See*
  Colonial dwarf-
  dandelion
Potentilla
  canadensis, **157**
  simplex, 157
Poverty rush. *See* Path rush
Prairie mimosa. *See*
  Bundleflower
Prickly lettuce, **166**
Prickly sow-thistle, **175**
Prickly woodrush, **391**
Privet
  border, 114
  Chinese, **114–15**
  common, 114, 115
  Japanese, 114
Prunella vulgaris, **274**
Prunus serotina, **362**
Pukeweed. *See* Indian
  tobacco
Purple coneflower, **212**
Purple loosestrife, **225–26**
Purple-node joe-pye weed,
  213
Purple-rocket, **263**
Purslane speedwell, 250

Quaker bonnet. *See* Spring
  bluets
Quaker ladies. *See* Spring
  bluets
Queen Anne's lace. *See* Wild
  carrot
Queen-of-the-meadow. *See*
  Purple-node joe-pye
  weed
Quercus
  alba, **344–45**
  macrocarpa, **346**
  marilandica, **347**
  montana, **348–49**
  muhlenbergii, 348
  rubra, **350–51**
  shumardii, 350–51
  stellata, 344, 345
  velutina, 350

Ragged fringed orchid, **116**
Ragweed
  common, 300
  giant, **300–301**
Ramp, 82
Ranunculus
  abortivus, 154
  ficaria, **152–53**
  micranthus, **154**
  recurvatus, **155**
  sardous, **192**
Ratibida pinnata, **168**
Rattlebox. *See* Square-pod
  water-primrose
Rattlesnake-master,
  87
Reclining St. Andrew's
  cross. *See* St. Andrew's
  cross
Redbud, **342**
Red clover, 109
Red columbine, **206**
Red dead-nettle, **203–4**
Red elm, **369–70**
Red maple, 379
Red mulberry, 357, 358
Redroot pigweed. *See* Rough
  pigweed

Rhamnus
  cathartica, **382–83**
  citrifolia, 382
  davurica, 382
Rhombic copperleaf, **310**
Rhus
  copallina, **296–97**
  glabra, 296, 297
  radicans, **298**
River oats. *See* River wood
  oats
River wood oats, **395**
Roanoke bells. *See* Virginia
  bluebells
Robinia pseudoacacia, **371**
Rock bells. *See* Red
  columbine
Rock chestnut oak. *See*
  Chestnut oak
Rock oak. *See* Chestnut oak
Rorippa sylvestris, **144**
Rosa multiflora, **128**
Rose-mallow, **111–12**
Rough buttonweed, 129
Rough pigweed, 294, 295
Rough sowthistle. *See* Prickly
  sow-thistle
Roundleaf ragwort. *See*
  Running groundsel
Round-leaved bittersweet. *See*
  Asian bittersweet
Roving Charlie. *See*
  Ground-ivy
Rubus
  alleghaniensis, **75**
  occidentalis, 75
Rudbeckia
  hirta, 170
  laciniata, **169**
  triloba, **170–71**
Rue-anemone, **72**, 73
Ruellia
  caroliniensis, 255
  strepens, **255**
Rumex
  altissimus, **316**
  crispus, **290–91**
  obtusifolius, 290–91

Run-away-Nell. *See*
Ground-ivy
Running buffalo clover, 18
Running five-fingers, **157**
Running groundsel, **141**
Running-myrtle. *See*
Common periwinkle

Sabatia angularis, **221**
Sage
lyreleaf. *See* Wild sage
wild, **242**
wood. *See* Canada
germander
Sagittaria latifolia, **81**
Sailor's tobacco. *See*
Mugwort
St. Andrew's cross, **184**
St. John's-wort
common, **179**
spotted, 179
Salix exigua, **365**
Salvia lyrata, **242**
Sambucus canadensis, **99**
Sandbar willow, **365**
Sandvine, **89**
Sanguinaria canadensis, **71**
Sanicula
gregaria, **299**
odorata, **299**
Sassafras, **353**
Sassafras albidum, **353**
Satin flower. *See* Honesty
Satin-flower. *See* Narrowleaf
blue-eyed grass
Saururus cernuus, **131**
Scaly-barked hickory. *See*
Shagbark hickory
Scilla siberica, **237**
Scrophularia marilandica,
231
Scutellaria
lateriflora, 224
nervosa, **278**
Sedum
acre, 65
ternatum, 65
Self-heal, **274**

Senecio
glabellus, **140**
obovatus, **141**
Senna hebecarpa, **182**
Sensitive fern, **36**
Sericocarpus linifolius, **94**
Serviceberry. *See* Downy
serviceberry
Sessile blazing-star, **214–15**
Setaria
faberi, 403
glauca, **403**
pumila, **403**
Sevenbark. *See* Wild
hydrangea
Shadblow. *See* Downy
serviceberry
Shadbush. *See* Downy
serviceberry
Shagbark hickory, **374**
Shaggy soldier. *See* Common
quickweed
Sharp-lobed hepatica, **247**
Shattercane, 407
Shellbark hickory. *See*
Shagbark hickory
Shepherd's purse, **52**
Shiny bedstraw, **130**
Short-fruited whitlow-grass,
55, 56
Shumard oak, 350–51
Siberian squill, **237**
Sicyos angulatus, **309**
Silphium perfoliatum,
172
Silver dollar plant. *See*
Honesty
Silver maple, **378**
Silver-rod, 174
Sisyrinchium angustifolium,
271
Small bluet, 248
Small-headed aster, 138
Smartweed
dock-leaved, **119**
dotted, **120**
Pennsylvania, **228–29**
Smilacina racemosa, **63**

Smilax
hispida, 318
rotundifolia, **318**
Smooth aster, **282**
Smoothbark hickory. *See*
Pignut hickory
Smooth dock. *See* Pale dock
Smooth false foxglove, **230**
Smooth goldenrod, 198
Smooth hedge-nettle, 223
Smooth pigweed, **294–95**
Smooth rose-mallow, 111
Smooth sickle-pod, **51**
Smooth Solomon's-seal, 63,
64
Smooth sumac, 296, 297
Smooth trailing lespedeza,
220
Smooth yellow violet, **159**
Snakeroot. *See* White
baneberry
Snowdrops. *See* Star-of-
Bethlehem
Soft agrimonia, **193**
Solanum
americanum, 279
carolinense, **279–80**
dulcamara, **281**
ptycanthemum, 279
Solidago
bicolor, 174
caesia, **197**
canadensis, **198–99**
flexicaulis, 173
gigantea, 198
hispida, **174**
ulmifolia, **200**
Solomon's-seal
false, 63
smooth, 63, **64**
true, 63
Sonchus asper, **175**
Sorghastrum nutans, **404–5**
Sorghum, 407
Sorghum
bicolor, 407
halepense, **406–7**
Southern adder's-tongue, **40**

Southern beech fern, **41**
Southern bladder fern, **35**
Southern cane. *See* Giant cane
Spanish-needles, 161
Spear thistle. *See* Bull thistle
Speedwell
    birdseye, **250–51**
    common, 252
    purslane, 250
    thyme-leaved, **252**
Spicebush, **147**
Spiderwort
    Virginia, **267**
    wide-leaved, 267
Spiranthes vernalis, **117**
Spotted lady's thumb, 228,
    229
Spotted St. John's-wort, 179
Spreading chervil, **42**
Spring avens, **158**
Spring bluets, **248–49**
Spring ladies-tresses, **117**
Spring messenger. *See* Lesser
    celandine
Square-pod water-primrose,
    188
Squirrel-corn, **66**
Stachys tenuifolia, 223
St. Andrew's cross, **184**
Staphylea trifolia, **386**
Star-eyed grass. *See*
    Narrowleaf blue-eyed
    grass
Star-of-Bethlehem, **68**
Stellaria
    media, **61**
    pubera, **62**
Stinking buckeye. *See* Ohio
    buckeye
St. John's-wort
    common, **179**
    spotted, 179
Stone mint, **286**
Stoneroot. *See* Northern
    horse-balm
Strawberry
    Indian, **156**
    wild, 156

Strawberry bush. *See*
    Hearts-a-bursting-with-
    love
Straw-colored flat-sedge, **389**
Stylophorum diphyllum, **151**
Stylosanthes biflora, **183**
Sugar maple, **379–80**
Sumac
    smooth, 296, 297
    winged, **296–97**
Summer bluet. *See* Broad-
    leaved bluets
Summer grape, **323–24**
Summer-snowflake. *See*
    Star-of-Bethlehem
Sunflower
    ashy, **163**
    divaricate, **162**
    woodland. *See* Divaricate
    sunflower
Swamp hickory. *See* Pignut
    hickory
Swamp-lily. *See* Lizard's tail
Swamp milkweed, **209**
Sweetgum, **352**
Sweet potato, 103
Sweet William. *See* Forest
    phlox
Switch grass, **402**
Sycamore, **359**
Symphoricarpos orbiculatus,
    100
Symphyotrichum
    cordifolium, **283**
    laeve, **282**
    novae-angliae, **284**
    pilosum, **138**
    racemosum, 138

Tall anemone, **123**
Tall bellflower, **264**
Taraxacum officinale, **142**
Teucrium canadense, **223–24**
Thalictrum dioicum, **292**
Thalictrum thalictroides, 72
Thelypteris hexagonoptera, **41**
Thimbleweed. *See* Tall
    anemone

Thin-leaved coneflower. *See*
    Three-lobed coneflower
Thistle
    bull, **210–11**
    Canada, **258–59**
    musk, 210, 211
    spear. *See* Bull thistle
Thlaspi
    alliaceum, **57–58**
    perfoliatum, 57
Three-leaved sedum, **65**
Three-lobed coneflower,
    170–71
Thyme-leaved speedwell, **252**
Tick-trefoil
    few flowered, **270**
    hoary, **270**
    naked, 219
    panicled, **219**
Tigernut. *See* Chufa
Tilia americana, **366**
Tipularia discolor, **331**
Toadshade trillium, **207**
Toothwort
    broad-leaved, 54
    five-parted, **54**
Torilis arvensis, 42
Touch-me-not
    orange, **176–77**
    yellow, 176
Tovara virginiana, **121**
Toxicodendron radicans, **298**
Tradescantia
    subaspera, 267
    virginiana, **267**
Tree-of-heaven, **376–77**
Trifolium
    pratense, 109
    repens, **109**
    stoloniferum, 18
Trillium
    flexipes, **77**
    sessile, **207**
Trillium
    bent, 77
    toadshade, **207**
True Solomon's-seal, 63
Trumpet creeper, **178**

Tulip magnolia. *See* Tuliptree
Tulip poplar. *See* Tuliptree
Tuliptree, **354–55**
Turkey pea. *See* American
    groundnut
Turnip, **143**
    Indian. *See* Jack-in-the-
        pulpit
    pepper. *See* Jack-in-the-
        pulpit
Twin-leaf, **46**
Typha latifolia, **332–33**

Ulmus
    alata, **368**
    americana, 369
    rubra, **369–70**
Uniola latifolia, **395**
Upland blueberry. *See*
    Low-bush blueberry
Upland hickory. *See*
    Shagbark hickory
Uvularia
    grandiflora, **160**
    perfoliata, 160

Vaccinium
    pallidum, **105**
    stamineum, 105
Valerianella
    radiata, 78
    umbilicata, 78
Veined skullcap, **278**
Venus' pride. *See* Broad-
    leaved bluets
Verbascum
    blattaria, 194
    thapsus, **194–95**
Verbena
    lantana, 135
    officinalis, 135
    urticifolia, **135**
Verbesina
    alternifolia, **201**
    virginica, 201
Veronica
    officinalis, 252
    peregrina, 250

persica, **250–51**
serpyllifolia, 252
Vervain, 135
Viburnum acerifolium, **101**
Vinca minor, **232**
Viola
    pubescens, **159**
    rafinesquii, 79
    sororia, **253**
    striata, **80**
Violet
    common blue, **253**
    cream, **80**
    dogtooth. *See* Yellow
        fawn-lily
    green, **293**
    smooth yellow, **159**
Violet wood-sorrel, **243**
Virginia bluebells, **234**
Virginia buttonweed, **129**
Virginia copperleaf, 310
Virginia cowslip. *See* Virginia
    bluebells
Virginia-creeper, **325**
Virginia spiderwort, **267**
Virginia spring-beauty, **205**
Virgin's-bower, 125
Vitis aestivalis, **323–24**

Wahoo. *See* Common
    arrowhead
Walking fern, **34**
Wandering Taylor. *See*
    Moneywort
Water beech. *See* Ironwood
Water-cress, **95–96**
Water dock. *See* Pale dock
Water-dragon. *See* Lizard's tail
Water hyacinth, 276
Waterleaf
    appendaged. *See* Biennial
        waterleaf
    biennial, **238**
    hairy, 238
    maple-leaved, **69**
Water-primrose
    creeping, **189**
    square-pod, **188**

Weedy bittersweet, 305
White ash, **385**
White avens, **126**
White baneberry, **122**
White-birdseye, 61
White clover, **109**
White fawn-lily, **70**
White mulberry, **357–58**
White oak, **344–45**
White pine, **336**
White poplar. *See* Tuliptree
White snakeroot, **90**
White vervain, **135**
Whitewood. *See* Eastern
    cottonwood
White wood. *See* Sycamore
Whitlow-grass, **55–56**
Wide-leaved spider-wort, 267
Wild ageratum. *See*
    Mist-flower
Wild basil. *See* Canada
    germander
Wild bergamot, **222**
Wild black cherry, **362**
Wild black rum. *See* Wild
    black cherry
Wild carrot, **85–86**
Wild comfrey, **233**
Wild geranium, **202**
Wild ginger, **328**
Wild hyacinth, **236**
Wild hydrangea, **110**
Wild leek, 82
Wild mustard. *See* Turnip
Wild oats. *See* River wood oats
Wild opium. *See* Prickly
    lettuce
Wild sage, **242**
Wild stonecrop. *See*
    Three-leaved sedum
Wild strawberry, 156
Wild sweet potato vine, **103**
Wild yam, **104**
Winged elm, **368**
Winged sumac, **296–97**
Wingstem, **201**
Wintercreeper. *See* Climbing
    euonymus

Wire-grass. *See* Path rush
Woodland lettuce, **262**
Woodland stonecrop. *See*
    Three-leaved sedum
Woodland sunflower. *See*
    Divaricate sunflower
Wood-nettle, **320**
Wood poppy. *See* Celandine
    poppy
Wood sage. *See* Canada
    germander
Wood-sorrel
    common yellow, **149–50**
    Illinois, 149
    violet, **243**

Wreath goldenrod. *See*
    Axillary goldenrod

Yam-leaved clematis,
    125
Yellow adders-tongue. *See*
    Yellow fawn-lily
Yellow archangel. *See* Yellow
    lamium
Yellow coneflower, **168**
Yellow fawn-lily, **148**
Yellow flag iris, **185**
Yellow foxtail, **403**
Yellow ironweed. *See*
    Wingstem

Yellow lamium, **187**
Yellow nutsedge. *See* Chufa
Yellow poplar. *See* Tuliptree
Yellow poppy. *See* Celandine
    poppy
Yellow snowdrop. *See* Yellow
    fawn-lily
Yellow star-grass, **146**
Yellowtop, **140**
Yellow touch-me-not,
    176
Yellow trout-lily. *See* Yellow
    fawn-lily

Zigzag goldenrod, **173**